P9-BYJ-773

Weapons of Tomorrow

Weapons of Tomorrow

Brian Beckett

PLENUM PRESS · NEW YORK AND LONDON

Dedication

To My Son

Acknowledgments

Although it is impossible to list all the people and organizations who have helped in compiling this book, a few deserve a special mention. First of all I would like to thank Michael Orr of the Royal Military Academy, Sandhurst, and Julian Perry Robinson of the Science Policy Research Unit, University of Sussex, for their constructive comments. Peter Thompson of Scientology, London, generously supplied countless CIA and official military documents on chemical and biological warfare obtained at the Public Record Office, London, or under the US Freedom of Information Act, and Chloe Haslam of Frost & Sullivan kindly allowed me to look at the company's market studies of military lasers. Beth Burack of *The Progressive* is thanked for the efficiency by which she provided me with the magazine's controversial articles on the H-bomb.

The staff at the International Institute of Strategic Studies (ISS), London, and the Press Office of the American Embassy have been especially helpful in dealing with endless queries. The vast majority of the raw data on strategic missiles is based on the ISS 1982 edition of *Military Balance*. The Public Relations offices of the US 3rd Airforce, Mildenhall, and US Naval Headquarters, London, have provided pictures at a moment's notice. Conclusions and possible errors or omissions are the sole responsibility of the author.

The author and publishers are grateful to the following for permission to reproduce photographs:
Aerospatiale, Paris 11; Associated Press 24, 29, 30; Boeing Aerospace Company, Seattle 5, 7, 18; Bundesarchiv, Coblenz 19; Hugh W. Cowin 6; Crown Copyright, Ministry of Defence, London 31; Department of Defense, Washington, D.C. 14, 25; General Dynamics, California 20, 21; Department of Energy, Las Vegas 2; Imperial War Museum London 26; Lockheed Missiles & Space Co. 10; MARS, London 4, 5, 7, 8, 10, 11, 13, 14, 16, 17, 18, 19, 25, 26, 31; Martin Marietta Aerospace, Florida 16; Press Association, London 23; Rockwell International, California 4; Times Newspapers Limited/David Case 22; U.S.A.F. 1, 8, 12, 15; U.S. Army 17; U.S. Navy 3, 9, 13; James Webb, London 28; Wellcome Museum of Medical Science, London 27.

Library of Congress Cataloging in Publication Data

Beckett, Brian.
Weapons of tomorrow.

Includes bibliographical references and index.
1. Weapons systems. 2. Arms and armor. I. Title.
UF500.B43 1983 355.8′2 83-9575
ISBN 0-306-41383-3

First Plenum Printing, 1983

Plenum Press is a Division of
Plenum Publishing Corporation
233 Spring Street, New York, N.Y. 10013

First published in Great Britain by Orbis Publishing Limited, London, 1982

Printed in the United States of America

Contents

Introduction

One of mankind's greatest achievements in the twentieth century is the ability to destroy his entire race several times over. Nuclear, chemical and biological weapons have the power to kill millions, irrespective of age, military status or political sympathy. There are thousands of nuclear warheads in existence today, but only one can destroy an entire city, while only a few kilograms of certain chemical and biological agents are theoretically capable of depopulating the whole planet.

Until recently the threat of widespread destruction seemed to be receding. Biological weapons were outlawed in a 1972 treaty, progress towards a similar chemical treaty showed signs of progress and the spirit of détente appeared to be reducing the danger of nuclear war. Now, however, in face of what she sees as aggressive Russian expansionism coupled with a build-up in her conventional and nuclear forces, the USA has warned the USSR of an arms race 'it cannot win', commissioned a new generation of strategic missiles, ordered the long-delayed neutron bomb into production, and is on the verge of producing chemical nerve gas weapons for the first time since 1969. Only the status of biological arms remains unchanged and that is looking increasingly fragile in the light of American charges that Moscow has repeatedly violated the treaty. Crash research programmes have been instituted to develop laser technology so that it can destroy satellite-based missiles, and a treaty limiting anti-missile systems — one of détente's main achievements — is unlikely to survive the eighties in anything like its present form.

However, despite widespread discussion and frequent media coverage, the exact form and usage of nuclear, chemical and biological weapons are often overlooked, misunderstood, or bedevilled with misconceptions. Almost every published diagram of the workings of a nuclear weapon is inaccurate or shows 1945 technology. Delivery systems are detailed but with little thought to the political and strategic purposes they supposedly fulfil. Yet to understand the true nature of the threat posed by these weapons it is important to understand not only what they are but the strategic thinking behind them and the purposes for which they are intended. The strategy of deterrence is a complicated one and is built upon assumptions and hypotheses which then determine the commissioning and deployment of new weapons. For these reasons this book concentrates on present and future developments in nuclear, chemical and biological weapons, and not on the changes in conventional weaponry, which are well covered in other publications. Moreover, if a major war between East and West does come, it will certainly be nuclear with possible use of biological and chemical weapons, and it is these which threaten the widespread destruction of civilian populations.

In various forecasts I have assumed that American cruise and Pershing II missiles will be based in Europe as planned, but any number of factors could change this. The growing hostility towards nuclear weapons in Europe may produce their rejection and the Labour Party in Britain has already committed itself to unilateral disarmament. The present negotiations between the United States and the Soviet Union may limit the numbers deployed or even ban them altogether. But even if these events occur, European arms limitation would only marginally effect the overall picture and would only push the nuclear threat to Europe back rather than eliminate it altogether. Nuclear-free zones are not immune to the effects of nuclear war, and the balance of peace worldwide is held by two increasingly mistrustful superpowers coupled with a growing number of smaller nations joining the nuclear club.

1 Nuclear and Thermonuclear Weaponry

The Basics of Nuclear Energy

In the first part of this century, Albert Einstein advanced the then revolutionary idea that mass and energy are ultimately the same thing. The formal statement of this relationship is the now classic $E = mc^2$ where m is the mass and c is the velocity of light in free space (299,792 kilometres per second). By the formula, a small amount of mass is equivalent to an enormous amount of energy but no current technology is capable of making the complete conversion. The atomic bomb converts just under 0.1 per cent of its reacting material into energy, and even the hydrogen bomb converts only about 1 per cent.

The key to nuclear energy lies in the neutron — an electrically neutral particle which, along with the positively charged proton, comprises the nuclei of all atoms heavier than hydrogen. Every element has a unique number of protons giving it a specific atomic number, but the number of neutrons is not fixed and each element may take different forms. Varieties of any element are called its isotopes. Each element thus has a mass number as well as an atomic number; the mass number is the total number of neutrons and protons in the nucleus. A nucleus of ordinary hydrogen, for example, consists of a single proton but its first isotope (deuterium) has one proton plus one neutron giving it a mass number of two. The second isotope of hydrogen (tritium) has two neutrons in the nucleus for a mass number of three.

Tritium, however, is radioactive; it decays into helium-3 over a half-life of 12 years by giving off beta radiation. A half-life is the time it takes for half the atoms in an unstable element to decay, so half a freshly created sample of tritium will have converted itself into helium-3 after 12 years, three-quarters after 24 years and so on until the process is, for all practical purposes, complete. Helium-3 happens to be stable, but there are many important cases where radioactive elements decay into something which is itself unstable and undergoes further radioactive decay until the chain finally culminates in a stable element.

Because it lacks a charge, the neutron can easily penetrate the massive electrical barrier surrounding an atomic nucleus and cause a reaction by colliding with it. What will happen depends upon a number of variables including the element concerned and the velocity or energy of the neutron. Neutrons have a wide range of velocities running from 'slow', which is measured in a few electron volts (eV), to extremely energetic, high velocities of more than 10 million electron volts. The electron volt (see note 1) is the basic measure for quantifying atomic energy and is often quoted in units of a million, MeV. The energy expended in lifting a 0.3 kilogram (¾ pound) weight 0.3 metres (12 inches) is equivalent to some 6.24 trillion million electron volts.

During the 1930s a number of experiments involving the bombardment of uranium with neutrons revealed mysterious traces of other elements in the débris. In 1939 physicist Otto Frisch identified them: upon absorbing a neutron, a uranium nucleus may break into two pieces much the way a drop of water can elongate and split in half. For the briefest instant of time, the uranium atom absorbs the extra neutron and converts into a higher isotope of itself but, being highly unstable, immediately begins to divide, in a process known as fission.

Natural uranium comes in two forms: uranium-238 and a rare isotope, uranium-235, which comprises only some 0.72 per cent of the uranium found in nature. Within a few months of Frisch's discovery it was shown that it was the rarer uranium-235 which fissioned after absorbing a neutron of ordinary to moderate velocity. As it happens, uranium-238 does fission but only after being struck by highly energetic neutrons in excess of 1 MeV.

There is no single outcome for uranium fission. The daughter elements (known usually as the fission products) may be any of several in the middle part of the periodic table — barium, krypton and bromine are fairly typical examples. But what always happens is that the fission products have a combined mass which is less than that of the original nucleus. The difference is only about 0.1 per cent but what is released is energy, in this case *c.* 200 MeV in each uranium fission. The complete fissioning of a single gram of uranium-235 would release the energy equivalent of a 1 million watt (megawatt) power station operating for just under a day.

Fission energy is released in two stages: prompt and delayed. The nucleus divides into two or three daughter nuclei and they begin to radiate neutrons, X-rays, gamma radiation and alpha particles — gamma radiation consists of very short wave-length, energetic X-rays, and alpha particles are nuclei of helium-4, the second element in the periodic table. The typical energy released by each uranium fission is 191 MeV, and the total time involved is only about one one-hundred billionth of a second. This ends the prompt phase. During the delayed phase, the fission products may emit further neutrons, some gamma radiation, considerable beta radiation (high velocity electrons) and about 12 MeV of anti-neutrinos, which are elementary particles with little or no mass and no electrical charge. The energy released during the delayed phase totals some 27 MeV.

Even after the delayed phase is essentially complete, the two daughter nuclei remain unstable and enter into a series of radioactive decays which finally culminates in a stable element. Fission products and their descendants in the decay chain are some of the most deadly substances known to man and form the principal constituent of radioactive fallout and nuclear waste.

The two to three neutrons released during fission make it possible to produce nuclear energy through a chain reaction which sustains itself. Not all the (average) 2.5 neutrons produced in one uranium fission will necessarily cause another fission. Instead they may be simply absorbed by another nucleus without splitting it or pass through the material and escape through the surface.

So for a chain reaction to be sustained, each fissioning nucleus must cause at least one further fission because if it fails to do so the reaction will quickly work itself out. But if each fission produces significantly more than one additional fission, the numbers of individual fissions will increase with every step of the chain and release a very large amount of energy within a very short time. This is the basis of a nuclear bomb.

The key to nuclear energy lies in a simple idea known as the *k* factor. This is the rate of neutron production per successive fission generation and is found by dividing the number of neutrons in any one generation by the number in the immediately preceding generation. Thus, if each fission generation produces as many neutrons as it releases, the reaction is self-sustaining but controlled. But if each fission generation produces twice as many neutrons as it releases (*k* equals 2), the reaction is uncontrolled as the number of neutrons is increasing geometrically. Where *k* is 2, the injection of a single initial neutron will cause one fission which will produce two additional fissions which, in turn, will cause four and so on in a geometric progression until the material is no longer able to contain the tremendous energies released within its core and an explosion erupts outwards. The time between fission generations is in the order of one 100-millionth of a second so it would take only some 80 millionths of a second for the 80 to 81 generations required to fission the 1.56 trillion trillion uranium atoms responsible for the destruction of Hiroshima.

Unlike uranium-235, uranium-238 will not fission when it is struck by low-energy neutrons. Instead, there is a high probability that the neutron will be captured, so converting the uranium-238 into uranium-239 by virtue of the extra neutron. Uranium-239, however, is unstable. It decays through the emission of beta radiation and has a half-life of just over 23 minutes. The resulting element, neptunium-239, has an atomic number of 93 and is described as being transuranic. 'Transuranic' literally means 'beyond uranium'; uranium, with 92 protons in its nucleus, was the heaviest element known until neptunium was created artificially at the University of California, Berkeley, in 1940. But neptunium-239 is also unstable (its half-life is just over two days) and undergoes further beta decay until it converts itself into a second transuranic element with an atomic number of 94. This is plutonium-239 which is also unstable but has an exceedingly workable half-life of 24,000 years.

Plutonium-239 is almost twice as fissionable as uranium-235 and may be fissioned by fast as well as slow neutrons. It also releases an average of three neutrons per fission compared to 2.5 for uranium, making it a far more efficient weapon material. Plutonium is also a potent emitter of highly toxic alpha radiation, which is comparatively harmless externally but exceedingly lethal if ingested or inhaled, and a few millionths of a gram of plutonium will prove fatal if it enters the body. In the early 1940s the world's complete supply of plutonium weighed only a few micrograms but the recent proliferation of nuclear reactors requiring large amounts of uranium-238 has created fairly abundant supplies.

The atoms of some unstable elements become sufficiently excited to divide on their own without absorbing a particle from an outside source — a process known as spontaneous fission. Plutonium — especially isotopes of plutonium-239 — has a high rate of spontaneous fission compared to uranium. This gives it a 'neutron background', a fairly constant level of fission activity which makes plutonium comparatively difficult to work into a practicable and safe bomb design. While 'weapons-grade' plutonium usually refers to pure or nearly pure plutonium-239 'reactor-grade' is used to describe material containing at least 20 per cent of other marginally less efficient plutonium isotopes, such as plutonium-240 and 242.

However, it is a mistake to imagine that reactor-grade plutonium cannot be adapted into a weapon. For years companies and governments keen to export nuclear reactors throughout the world have understated reactor plutonium's weapons potential. All plutonium isotopes are fissionable and, although reactor-grade material imposes problems, there is no reason why competent engineering cannot solve them; the point was proved in 1977 when the United States exploded a bomb based on reactor-grade plutonium.

Nuclear Weapon Design

Building a nuclear weapon from either metallic uranium or plutonium involves a number of variables which centre upon the idea of a critical mass. A critical mass is the smallest amount of fissionable material required for a self-sustaining nuclear reaction, but it is not a fixed quantity. It varies according to the type of material used, its shape, and the design particulars. When a neutron is released through fission, it has to travel a certain distance before it fissions another nucleus. This distance is determined by such things as density and neutron energy, and is usually known as the 'mean time between fissions'. For uranium or plutonium at their normal densities, there is a size where a self-sustaining reaction is (for all practical purposes) certain, but below which it will not occur. The size is not fixed, however, because there are a number of ways to increase the neutron multiplication rate.

First of all, the shape is important. The larger the surface area of the nuclear mass, the more room for neutrons to escape outwards. Thus a sphere is the most efficient geometry because the ratio of surface area to volume is the smallest. Secondly, if those neutrons which do escape can be sent back into the nuclear material, fission efficiency is increased. To achieve this, the nuclear material is surrounded by a neutron reflector. Shape and reflection reduce the amount of material required to sustain a chain reaction of fissioning atoms. A 15-kilogram (33-pound) sphere of pure uranium-235 enclosed in a reflector made of uranium-238 is critical, but the critical masses for spherical uranium-235 are 28 kilograms (61 pounds) if enclosed in a reflector of aluminium, 20 kilograms (44 pounds) for a reflector of nickel and 11 kilograms (26 pounds) if the reflector is made of beryllium.

The purity of the nuclear material is also important. The presence of uranium or plutonium isotopes increases the total amount of material required, as the table on page 12 shows. Increasing the proportion of uranium-235 to

	Fissionable Isotope	Percentage Enrichment	Metalic Density (grams per cubic centimetre)	Critical Mass (kg)
Spherical Metal — Bare	Uranium-233	99	18.2	16.2
Spherical Metal — Bare	Uranium-235	95	18.5	51.3
Spherical Metal — Bare	Plutonium-239[2]	95	15.6	16.3
Spherical Metal — Bare	Plutonium-239[2]	100	19.8	10
Spherical Metal — Reflected	Uranium-233	100	18.2	5
Spherical Metal — Reflected	Uranium-235	100	18.5	15
		80		21
		60		37
		40		75
Spherical Metal — Reflected	Plutonium-239	100	19.8	4.4
		90		5
		80		5.6
		70		6.7
		60		7.8
		50		9.6

Fig. 1 Critical Mass

1. Thick uranium-238 reflector. Beryllium reflectors enable smaller critical masses; 11kg and about 4kg for pure U-235 and Pu-239 respectively.
2. Plutonium has various densities ranging from 15.6 to 19.8 grams per cubic centimetre.

uranium-238 — or plutonium-239 to any of its isotopes — is called enrichment. Reactor-grade uranium may be enriched by only a very few percentage points but weapons-grade generally refers to concentrations of uranium-235 of 93 to 99 per cent, although 50 per cent is adequate (just). Plutonium is more complex because most of its isotopes are fissionable. A critical mass of plutonium is increased by 77.2 per cent if the proportion of plutonium-239 is changed from 100 to 60 per cent; a corresponding reduction in the purity of uranium-235, however, will increase the critical mass by 146.6 per cent. This comparative insensitivity to changes in isotopic impurity is one of the principal reasons why most reactor-grade plutonium is considered viable weapon material. In a thick uranium-238 reflector, it takes just under 10 kilograms (22 pounds) of spherical plutonium to make a critical mass with only 50 per cent plutonium-239.

To make an efficient nuclear weapon, it is necessary to create a very high rate of neutron multiplication as rapidly as possible rather than simply a self-sustaining controlled reaction; in other words, the instant formation of a 'supercritical' mass with an escalating chain reaction from previously safe material. There are two chief methods of doing this. The first and best known is the 'gun' method, as used in the uranium bomb dropped on Hiroshima. Two pieces of uranium-235 are separated at opposite ends of what amounts to the

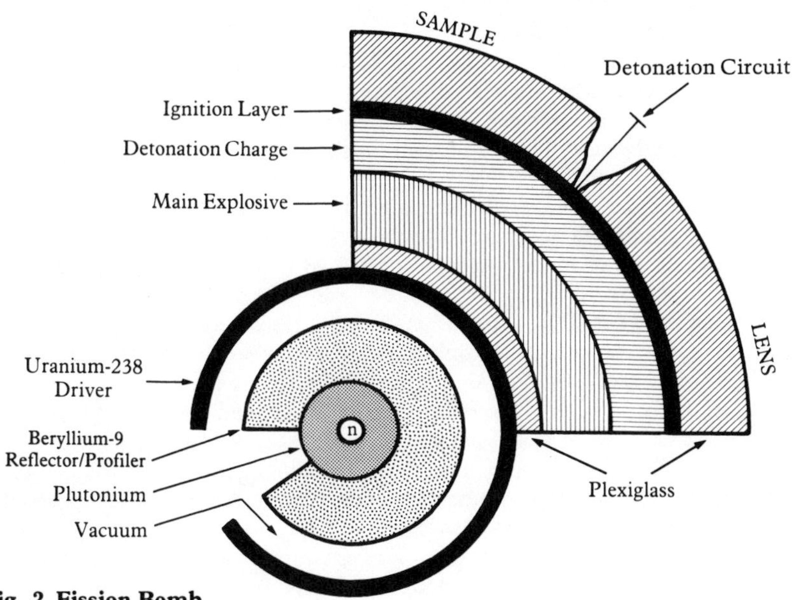

Fig. 2 Fission Bomb

n is the neutron source (fission initiator) which usually consists of deuterium and tritium.

barrel of a gun. Neither separately is large enough for a critical mass but when brought together they form a single piece which is supercritical. One piece is usually larger than the other and is fixed in place within a shell of neutron-reflecting material. At the other end of the barrel is the smaller piece, which is backed by a layer of conventional explosive that is linked to a detonating circuit. Upon detonation, the circuit closes, fires a primer and ignites the explosive which then propels the uranium down the barrel to join and fit with the larger unit at a speed — in the case of the Hiroshima bomb — of about 300 metres per second (670 miles per hour).

Because of its higher rate of spontaneous fission, plutonium cannot be used in the gun method because, at that assembly speed, the two pieces of plutonium are very likely to react prematurely and lead to what is known as a 'fizzile yield'. Premature fission may cause the material to heat and alter its shape or even melt, making an effective explosion impossible. Even if this does not happen, the energies generated in premature fission will force the materials apart before they have had time to react sufficiently. Instead, plutonium is detonated by uniformly compressing a sub-critical piece to a supercritical density at a speed that avoids a destructive level of such premature fission.

Uranium may also be exploded in this manner. Increasing the density reduces the mean distance and time between fissions and lowers the number of escaping neutrons by decreasing the surface area. Compression is accomplished by detonating a configuration of advanced but conventional high explosives — triamino trinitro benzene, for example — to generate a uniform shock wave which implodes the fission material to densities several times normal and at speeds up to 5 kilometres per second or more. At this sort of assembly speed high reactivity is created in the plutonium core before spontaneous fission can cause premature detonation.

Arrangements of high explosives designed to direct the majority of their energy in a specific direction are known as 'shaped charges' and have found a wide range of use in modern weaponry. When they are designed to cause an imploding, compressive shock wave, they are often called 'lenses' because of their focusing effect. The illustration on page 13 sketches a highly advanced and previously unreported plutonium bomb with a soccer-ball-like configuration of 32 explosive lenses divided into an interlocking pattern of 20 hexagons and 12 pentagons. Simultaneous detonation of each lens is vital; otherwise the result is fizzile yield or a conventional explosion spreading radioactive débris and gases. When the lenses fire, a spherically symmetrical shock wave converges on the heavy metal liner/driver and accelerates it to a velocity up to 10 kilometres per second, or even more — a speed far greater than that of a rocket escaping the earth's gravity. The vacuum layer allows the driver to accelerate before impact. On striking the light metal reflector, the driver's momentum creates a shock wave which converges on the core. When the shock wave meets the core, two more shock waves are created. One propagates inwards to begin compression but the second moves backwards to strike the driver where it is reflected back into the collapsing plutonium.

Careful design maximizes the reflecting effect and aims at creating a series of converging shock waves which overtake each other and converge at the core's centre. Known as a 'tailored' or 'profiled' explosion, this technique enables designers to obtain extreme and rapid compressions which allow highly efficient weapons from relatively small amounts of plutonium or uranium. The key is a 'soft' reflector. As well as returning escaping neutrons into the fissioning centre, the light reflector acts as an explosion 'profiler' between the heavy driver and core. Its presence allows the designer to tailor the converging shock waves for maximum compression. For selected materials, core size and driver velocity, there is an optimal reflector thickness which determines maximum pressure. Increasing the driver's velocity and arranging the reflector/profiler into concentric layers adds to the compression. When beryllium is used, its neutron reflecting properties increase as it is pressed into a super-density between the driver and the plutonium. As the core begins to erupt, the inertia of the reflector and driver 'holds' it to a supercritical density for an extra fraction of a second. This allows further fissions before it expands to a sub-critical size and — because the fissions are increasing geometrically — the yield gained is dramatic.

Using a far more primitive design than the above, the Nagasaki bomb was about 10 times as efficient as the Hiroshima gun weapon in terms of yield per unit of production cost. The efficiency of the implosion method has remained superior despite improvements to both. However, the first implosion assemblies were far more delicate and weapons designed to resist tremendous shock may still use the gun design. Nuclear artillery shells, for example, must withstand the intense inertia imparted to the projectile when it is fired. Weapons designed to penetrate the ground before detonation (earth penetrators) must survive impact. The extent to which new implosion assemblies have been engineered to resist shocks of these magnitudes is not known but, if their delicacy is still a constraint, weapons of this sort are likely to be the only remaining examples of the gun approach.

Thermonuclear Weapons

In 1942, some six months after Pearl Harbor, J. Robert Oppenheimer, recently appointed head of design for the Manhattan Project, crossed the United States for an urgent consultation with one of his superiors. What so disturbed Oppenheimer was a number of theoretical calculations suggesting that the intense energies released by a successful nuclear bomb could be used to trigger far more powerful reactions in deuterium and tritium, the two isotopes of hydrogen, by causing them to convert into helium. The theory was based on a phenomenon known as fusion, which is the opposite of fission.

Under certain conditions of intense temperature and pressure, two atomic nuclei may combine to form larger elements. With the very light elements — hydrogen, helium, lithium and so on — the reactions are generally exothermic (in other words, they give off energy in the process). When the new nucleus has less mass than the two parents, the difference is released as energy. Although these fusion reactions give off far less energy that the fission of uranium or plutonium, on a weight-for-weight basis many more atoms are involved, and the size of fusion (hydrogen) weapons are dictated only by economic and practical military considerations.

In a sense, fusion is the basis of all life because it is the means by which the stars, including our Sun, produce their energy. But stellar processes involve fusion chains which are more complicated than the reactions discussed here. In order for fusion to occur within the Sun's interior or a hydrogen bomb, the material must be compressed and heated to tremendous temperatures in order to give the two positively charged nuclei sufficient energy to overcome the electrical repulsion that exists between them. Two deutrons (deuterium nuclei), for example, must approach each other within a distance of some three quadrillionths of a metre before their fusion can occur and the temperatures required begin at between 10 million and 20 million degrees Centigrade. After this, the proportion of fusion increases dramatically with temperature until it becomes virtually complete at 100 million degrees Centigrade and beyond. Long before there is a significant chance of fusion, the individual atoms have been ionized or stripped of their electrons so creating the fourth state of matter — a 'plasma' consisting of rapidly and randomly moving electrons and positively charged nuclei. Because they require such intense temperatures,

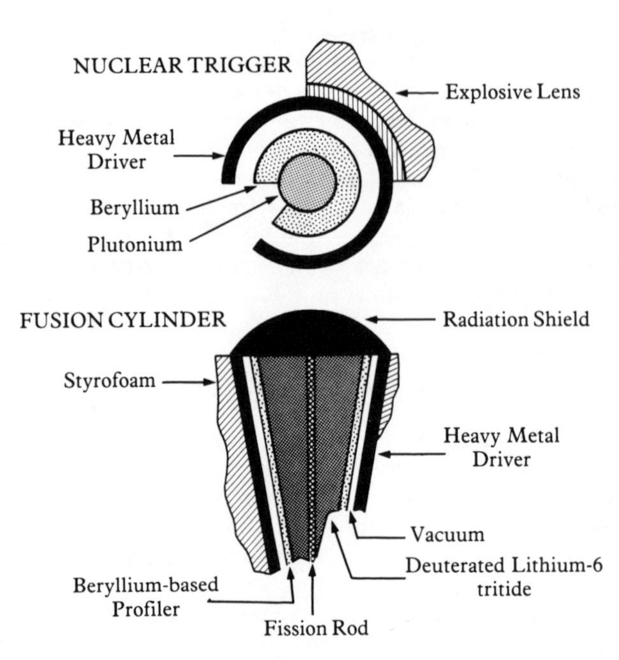

Fig. 3 The Teller-Ulam H-Bomb

X and gamma radiation from the trigger reflects off specially designed inner casing walls and explodes the styrofoam. This implodes the cylinder and causes fusion.

The particulars illustrate the idea of a neutron (enhanced-radiation) bomb rather than a weapon aimed for maximum blast and fallout.

fusion reactions are 'thermonuclear' and a hydrogen bomb is more accurately described as a thermonuclear weapon.

Basic research into the possibility of hydrogen weapons was carried out during World War II but it was not until after the Russians exploded their own atomic bomb in 1949 that the United States committed itself to a full-scale effort to develop them. The chief problem was to find a way of efficiently using a nuclear bomb to trigger the fusion device before the explosion destroyed it. The answer was found by the physicists Stanley Ulam and Edward Teller and has been known as the Teller-Ulam configuration ever since. But that configuration remained an official secret until 1979 when two American magazines independently published articles which revealed its essentials. One, the liberal journal *The Progressive*, scheduled an article for April 1979 outlining H-bomb design and was promptly taken to court by the Department of Energy which obtained an order restraining publication on the grounds of national security.

In the same month *Fusion*, a specialist magazine, published an entire issue devoted to some of the essential concepts behind thermonuclear theory and technology. *Fusion*'s aim was to show that governmental secrecy about basic physical principles, simply because they had military applications as well, was hindering scientific progress. Curiously, and to the magazine's surprise, the government ignored *Fusion* and concentrated its efforts on fighting *The Progressive* for six months, then dropped its case. When it was all over, it was clear that there had been no secret at all. *The Progressive*'s author, untrained in physics by his own admission, had obtained all his material from open sources or interviews, and Teller himself had revealed one of the main essentials in an *Encyclopedia Americana* article which had been on library shelves for the previous 10 years.

Hydrogen weapons are triggered by an implosion fission device as described earlier. About 300 millimetres (12 inches) or so away from the nuclear trigger is the fusion device, a tapering metal cylinder filled with the chemical compound lithium-6 deuteride. This compound is a grey crystalline material which is thermally stable up to a melting point of 680°C but highly unstable in the presence of air, when it reacts to moisture and decomposes. It is produced by passing molten lithium through a vessel filled with deuterium gas and, when used as a weapons fuel, heated to a ceramic consistency. Lithium-6 tritide, a mixture of lithium and tritium, is prepared in an identical manner and also reacts adversely with air. In most cases there is a rod of enriched fission material running down the centre of the cylinder to act as a second fission trigger. The cylinder itself is embedded in a dense polystyrene-type styrofoam at one end of the bomb-casing. At the other end is the fission bomb. The weapon casing itself is a cylindrical shell of intricately machined metal layers designed to focus the X-rays and gamma radiation emanating from the exploding fission trigger into the styrofoam encasing the fusion cylinder. At the top of the cylinder there is a shield to protect it from direct exposure to radiation.

All electromagnetic radiation travels at the speed of light and is composed of elementary particles known as photons. Photons lack mass but have certain important characteristics in common with conventional particles — they have momentum, for example. Leaving the fissioning trigger at the speed of light, the X-ray and gamma photons travel some 300 times faster than the expanding nuclear explosion and are able to bring about fusion in the lithium-6 deuteride before it is engulfed and destroyed. Reflected from the bomb-casing wall, these photons smash into the styrofoam, instantly converting it into an energetic plasma which implodes the fusion cylinder uniformly and compresses the lithium-6 deuteride to about 10,000 times its normal density by creating a unique type of (isentropic) shock wave which propagates inwards without heating the fuel below its surface. As the shock wave converges on the cylinder's centre, the fission rod is compressed towards criticality whereupon it begins to burn and release tremendous amounts of heat and neutrons to fuel reactions in the lithium-6 deuteride, which has also been heated and converted into plasma in a 'burn' emanating from its surface. The principal fusion reactions which occur when ignition temperature is reached are listed in note 2.

Modern hydrogen weapons probably do not contain any tritium to begin with. Instead, the tritium is created by the splitting of nuclei of lithium-6 by neutrons released from the fissioning rod and in the reaction of deuterium with deuterium and then deuterium with tritium. The energies released in these first reactions rapidly increase the temperature, which fuels and accelerates further reactions between deuterium and deuterium, deuterium and tritium, and tritium with tritium. In addition, there are several secondary reactions involving deuterium, tritium and helium-3.

If the casing of the fusion cylinder is made of uranium-238, the high velocity neutrons produced inside will cause a large number of fissions when they strike its surface. This both adds greatly to the overall energy yield of the bomb and further accelerates the fusion process within the cylinder. But the technique also adds considerably to the level of fallout by the creation of further radioactive fission by-products, and so-called 'clean' bombs do not include uranium-238 in the casing materials. Those which do, however, are nuclear weapons as much as they are thermonuclear ones because they run through a chain going from fission to fusion and back to fission. The exact proportion of fission within any thermonuclear weapon is secret, but in 'clean' devices which eliminate the third fission stage it is a small percentage. In 'dirty' weapons, on the other hand, it may be as much as 50 per cent, or a great deal more if the aim is deliberately to increase environmental contamination through fallout.

Fission-Fusion Hybrids and Mini-Nukes

Weapons-grade plutonium and uranium require a source of neutrons to initiate the diverging chain reaction. The earliest Manhattan Project versions apparently consisted of a small ball of alpha-radiating polonium-208 surrounded by a very thin shell of suitable alpha-shielding material. This was further coated with a layer of beryllium or lithium which give off neutrons when bombarded by alpha particles. When the ball was crushed in the formation of the supercritical mass, the polonium irradiated the outer material and neutrons flooded into the core.

Fusion-based sources produce far more neutrons than the Manhattan Project design and have been incorporated into nuclear weapons for years. They consist of a tiny balloon made of deuterated polyethylene or metallic foil and filled with pressurized deuterium and tritium gas. As the core collapses and the shock wave converges on its centre, the balloon is compressed causing copious neutron-producing fusion reactions. Another approach appears to be to use lithium deuteride as a reflecting material and explosion profiler.

However, advanced implosion techniques have created a new generation of weapons, known as 'mini-nukes'. These are warheads with comparatively small yields inserted into a wide variety of weapons — including artillery shells, land mines, naval depth charges and torpedos and just about any type of traditionally conventional weaponry imaginable — in order to increase the killing effect of tactical weapons. Mini-nuked depth charges or torpedos, for example,

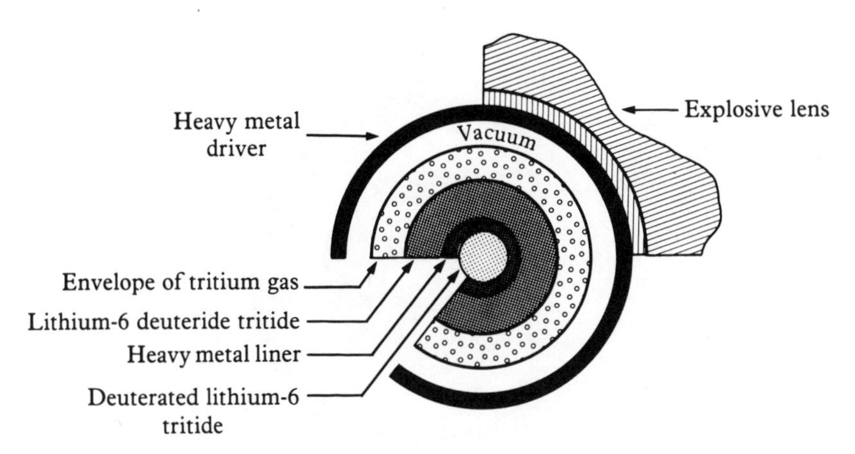

Fig. 4 Fission-Free Neutron Bomb

The fission-free H-bomb may already be a reality. An efficient driver could generate enough fusion energy to cause the explosive shock wave to compress the core to hyper-densities of several 10,000 times the normal. The liner, a heavy metal of variable density, would profile the shock wave for maximum compression.

Because of size considerations, yields would probably limited to a kiloton or so and, in such things as artillery shells, the device would probably be a cylinder rather than a sphere. This would be less efficient but, with high compression, the weapon should still be highly effective.

The particulars illustrate the idea for an enhanced-radiation, neutron weapon. The core might release around 900 tons of TNT equivalent with up to another 100 coming from the surrounding shell.

An example of a suitable liner might be a copper-beryllium alloy and the driver could be made of a dense metal such as iridium.

could yield the equivalent of several hundred tons of TNT without appreciably increasing in size. The technique required is to explode small amounts of fissionable material efficiently, and Polish scientific papers on tailored explosions report reductions in the critical mass of uranium-235 by over two orders of magnitude and compressions by factors of five to seven. Small cores have the additional advantage of being more easily compressible.

But there are other approaches to making mini-nukes. In 1977 Polish scientists reported the distinction of being the first to produce fusion from conventional explosives. A configuration of two cones surrounded by high-explosive was fired and generated an intense shock wave downwards into a small conical target filled with deuterium gas. The deuterium was compressed to 1000 times its initial density and significant fusion activity followed. Later the same scientists outlined an advanced 'bi-conical' configuration capable of compressing small amounts of uranium or plutonium into hyper-densities with

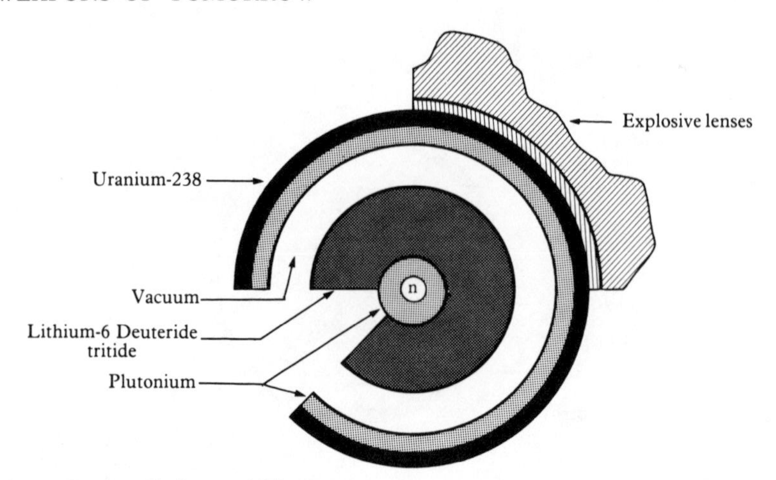

Fig. 5 Fusion-Enhanced Weapon

The two essentials are the heavy driver with an inner plutonium shell and the profiler/reflector of the low-density, light fusion-fuel. Accelerated to high speeds (10 or even 20 kilometres a second), the driver's impact on the fusion fuel creates both an intense and reflecting compressive shockwave on the plutonium core and fusion activity in the lithium deuteride.

Although considerable fusion energy is released, its chief purpose is to convert the 17.6 MeV and 3.26 MeV fusion neutrons into plutonium fissions releasing 200 MeV.

The weapon can probably be designed so that the plutonium shell goes supercritical as it collapses on the lithium deuteride. This would both increase the initial yield and, as it began to blow apart, it would expand inwards as well as outwards. The lithium deuteride would be effectively trapped between the expanding supercritical core and shell for a final spiralling boost in fission-enhancing fusion. The yield is boosted even further by fission in the uranium-238.

the added aid of explosion profiling. Such a system would make highly effective mini-nukes and a version of the basic design is illustrated on page 21. The detonation of the outer 'driver' cones causes the inner 'cumulative' cones to focus intensive (mach) shock waves on a small uranium or plutonium ball and compress it in a quasi spherical pattern to a hyper-density. Adapted into a weapon, the fission sphere would probably contain a micro-balloon neutron source at the centre.

The proportion of mini-nukes in the world's arsenals is not known but is certainly considerable, and it is likely that weapons laboratories have devised or will devise mini-nuke variants of conventional munitions without any real regard to their military practicalities or political desirability.

Future Nuclear Weapons

Few experts expect dramatic breakthroughs in nuclear weapon technology in the near future and certainly nothing approaching the quantum jump achieved

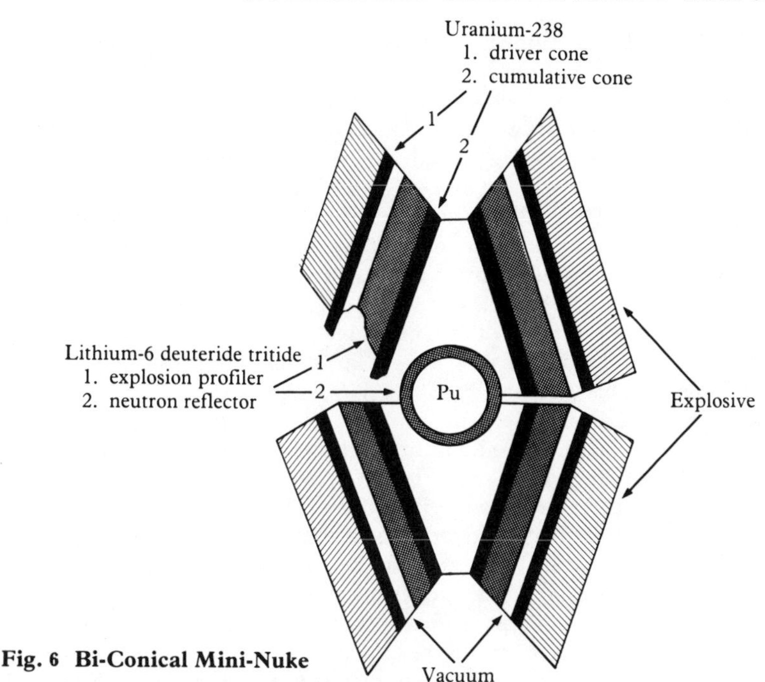

Fig. 6 Bi-Conical Mini-Nuke

The biconical system is said to be capable of compressing uranium or plutonium by values of five to seven and reducing critical mass to between 54 and 100 grams. If so, it would be possible to create high levels of supercriticality with masses of as little as 200 to 400 grams. The fission cores of such bombs would have diameters of only 2.682 to 3.38 centimetres (1.05 to 1.33 inches) but be capable of one or two kiloton yields. Very efficient weapons using fusion enhancing reflectors might enable reliable yields up to 4 or 5 kilotons. (Major reference: S. Kaliski, *Journal of Technical Physics*, Warsaw, vol. 19, 4, 1968.)

by fusion over fission. Work continues on mini-nuking, improving general efficiency and yield-to-weight ratios or the perfecting of tailored-effect warheads such as the enhanced-radiation, neutron bomb. Instead of looking for radical developments, now there is an increasing demand to extend the life-spans of existing weapons due to a relative scarcity of essential materials and a growing opposition to underground testing. New 'super' bombs surpassing hydrogen weapons must await advances in science over the next decade and probably far beyond. But there are still significant and disturbing possibilities for the long term within present-day capabilities.

First, there is a long-debated possibility of constructing extremely small weapons based upon the fissioning of the higher transuranic elements. At the moment there are 13 transuranics, ranging from neptunium to hahnium, which has an atomic number of 105. Many of these have little use outside the research laboratories where they are created in nuclear reactions, but a few are finding application in such things as medical technology and geological exploration.

Californium (element number 98) emits roughly 3.5 neutrons per fission and it may be possible to make a bullet-sized nuclear weapon of it, although nobody as yet seems to have given the matter much attention. One suggestion has been to build a supercritical mass of californium and include control elements which would absorb sufficient of the background neutrons caused by spontaneous fissions to keep it below the point of criticality. Boron and cadmium, for example, are highly efficient neutron absorbers and are used in reactor control rods. To detonate the weapon, the control device would be withdrawn and turn the californium supercritical. But, so far as anyone knows, these possibilities remain conjectural and there has been little serious work on developing the heavier transuranics into military implements — metallic californium, for example, has yet to be prepared. But there are advocates of these programmes and Dr Teller has been quoted as supporting such work.

Transuranic weapons offer the prospect of extremely small explosive packages. Missile warheads would be dramatically reduced in size and enable a single launcher to carry many more than at present. A single re-entry vehicle could even carry several warheads, giving ballistic missiles a sort of 'shotgun' effect and allowing an attacker to saturate an enormous area with hundreds of tiny but powerful nuclear weapons. Exactly what might tempt either side to begin work on such a programme is difficult to predict but the emerging likelihood of new ballistic missile defences and generations of mobile intercontinental ballistic missiles might well do so.

Another possibility is a fusion weapon that does not require a nuclear bomb as a trigger. Although all thermonuclear weapons produce some radioactive contamination (due to neutron irradiation of the environment), they are 'cleaner' than fission weapons because most products of fusion reactions are not themselves radioactive, or not dangerously so. Pure thermonuclear weapons, however, may be developed from current work on the commercial applications of fusion energy. Scientists have long been striving to make a viable fusion reactor but no-one has yet succeeded in developing a means of containing the plasma for the necessary length of time. At the temperature generated, any contact with the reactor wall would immediately destroy it. Two answers have been posited: magnetic confinement and inertial confinement. The former basically involves introducing a magnetic field between the plasma and the reactor's walls, but it involves long reaction times (up to a second) because the fuel is not highly compressed. Much of this work is Russian in origin, and their most promising design, the Tokamk, was developed in the USSR. Inertial confinement consists of imploding a deuterium-tritium pellet by irradiating it with laser light or accelerated ions and allow inertia to 'hold' the plasma together for the required time. Upon striking the pellet, the electrons create an isentropic shock wave and generate 'soft' X-rays to ionize its surface. Soft X-rays are a commonly known scientific phenomenon — they cause TV screens to emit X-rays, for example — and are an example of 'bremsstrahlung' radiation (German for 'braking' radiation). Braking radiation is caused by an energetic, electrically charged particle decelerating in matter.

In the fusion pellet, the soft X-rays ionize its surface but do not penetrate deeper to heat the fuel and make compression non-isentropic. X-rays will also damage delicate electronics and braking radiation is one of the chief mechanisms envisaged for a new generation of particle beam weapons now being researched with the aim of damaging a target by striking it with a beam of accelerated particles.

Fusion may now be generated in several ways. An intense electric current may be introduced into a pure gaseous mix of hydrogen isotopes, generating temperatures of 100 million degrees Centigrade. Polish work with explosive-generated fusion shows the possibility of future weapons on these lines but the practicalities are probably well beyond current capabilities.

The most immediate possibility for fission-free hydrogen weapons probably lies with the 'fast liner' — a technique combining both magnetic and inertial confinement. A cylinder filled with fusion fuel (the liner) is surrounded by a second cylinder linked to a capacitator bank. When fired, an electric current from the capacitators passes into the outer cylinder and, in turn, creates a second current in the liner flowing in the opposite direction, which produces a magnetic field with force enough to implode the liner. Within the liner, the magnetic field is compressed to more than 1000 times its normal strength and presses the plasma within it. Other approaches use lenses of conventional explosives to compress the magnetic field.

Progress in fast liners might offer countries a route to fusion weapons without the expense and complexities involved in the design and construction of effective nuclear triggers. Tritium must normally be made by bombarding lithium with neutrons in a reactor but deuterium is plentiful (comprising about 16 out of every 100,000 hydrogen atoms in nature). The ideal 'clean' fusion fuel would be a boron-11 hydride because neither the boron/proton reaction's precursors or its products (three alpha particles) is radioactive. This complicated (three-stage) reaction is one of the ultimate hopes of commercial fusion research but it requires temperatures and pressures beyond anything now obtainable.

Totally new types of nuclear weapon may one day be a reality but, so far, these seem more science fiction than anything else. The favourite is the anti-matter bomb, which does, in fact, have a theoretical basis in physics since all elementary particles have their 'anti' counterparts. The positron, for example, is an electron carrying a positive rather than a negative charge. When it collides with a 'normal' electron the two annihilate each other in a total mass-energy conversion released in gamma radiation. But the technology required to produce anti-matter in any quantity, much less adopt it into a workable weapons system is, as yet, unavailable and is likely to remain so for the foreseeable future. This is probably fortunate since the potential energy yields are beyond anything now obtainable. If a single gram of uranium-235 were totally converted into energy, for example, it would be the equivalent of two Hiroshima bombs. However, nuclear weapons introduced over the next decade are not likely to be qualitatively different so much as dramatically improved through more efficient designs or the adoption of new technologies.

2 Present-day Nuclear Weapons

Classifying Nuclear Weapons Systems

The intense temperatures generated by nuclear and thermonuclear weapons vaporize their casing materials, fission products and unused fission or fusion fuel into hot gases which are violently released and create a shock or blast wave in air, earth or water. This wave is distinct from those proportions of the yield expressed in thermal radiation or in the radiation of neutrons, gamma rays, beta and alpha particles and X-rays. A weapon's total yield is measured in either kilotons (thousands of tons of TNT equivalents) or megatons (thousands of kilotons). To get an idea of what these energy levels mean, the detonation of a 1 kiloton nuclear device releases the energy equivalent of a 1 megawatt electric power station's output for 48 days. The 12-kiloton bomb dropped on Hiroshima released the energy equivalent of 18 months' output while a 1 megaton thermonuclear explosion releases the equivalent of 132.5 years' production (see note 3).

The capacity of nuclear weapons to devastate cities lies principally in their blast effect, which is usually measured in pounds per square inch (psi) overpressure – that is to say, pressure greater than normal atmospheric. But the overpressures generated do not increase proportionately with increases in yield and so the urban damage potential of various yields cannot be directly compared with any real degree of accuracy. Comparison is done by means of a simple scale which converts yields into megaton equivalents (Mte); see note 4. Although the yield of the 24-megaton warhead carried on some Soviet missiles is 2000 times greater than the Hiroshima bomb, the difference in megaton equivalents is a factor of only 162 (see the graph on facing page). Megaton equivalence is only used to compare the respective damage potentials of various yields on populated areas and has no genuine military meaning in the sense of effectiveness on any battlefield or against any sort of military target.

Nuclear or thermonuclear weapons are usually classed as 'strategic', 'theatre' or 'tactical', but it is almost impossible to draw fine distinctions. 'Strategic' normally refers to delivery systems based in the United States and with intercontinental ranges and, in popular association, to fairly large yields. 'Theatre' generally refers to systems with ranges restricted to specific areas of operation such as between western and eastern Europe. 'Tactical' means much the same thing but usually refers to battlefield distances of a few hundred kilometres and normally much less. A phrase which has grown in popularity recently is 'Eurostrategic'. Generally, this refers to aircraft and ballistic or cruise missiles with ranges of 1600 to 3200 kilometres or so (1000 to 2000 miles); in other words, systems capable of operating between the western areas of the Soviet Union and western Europe.

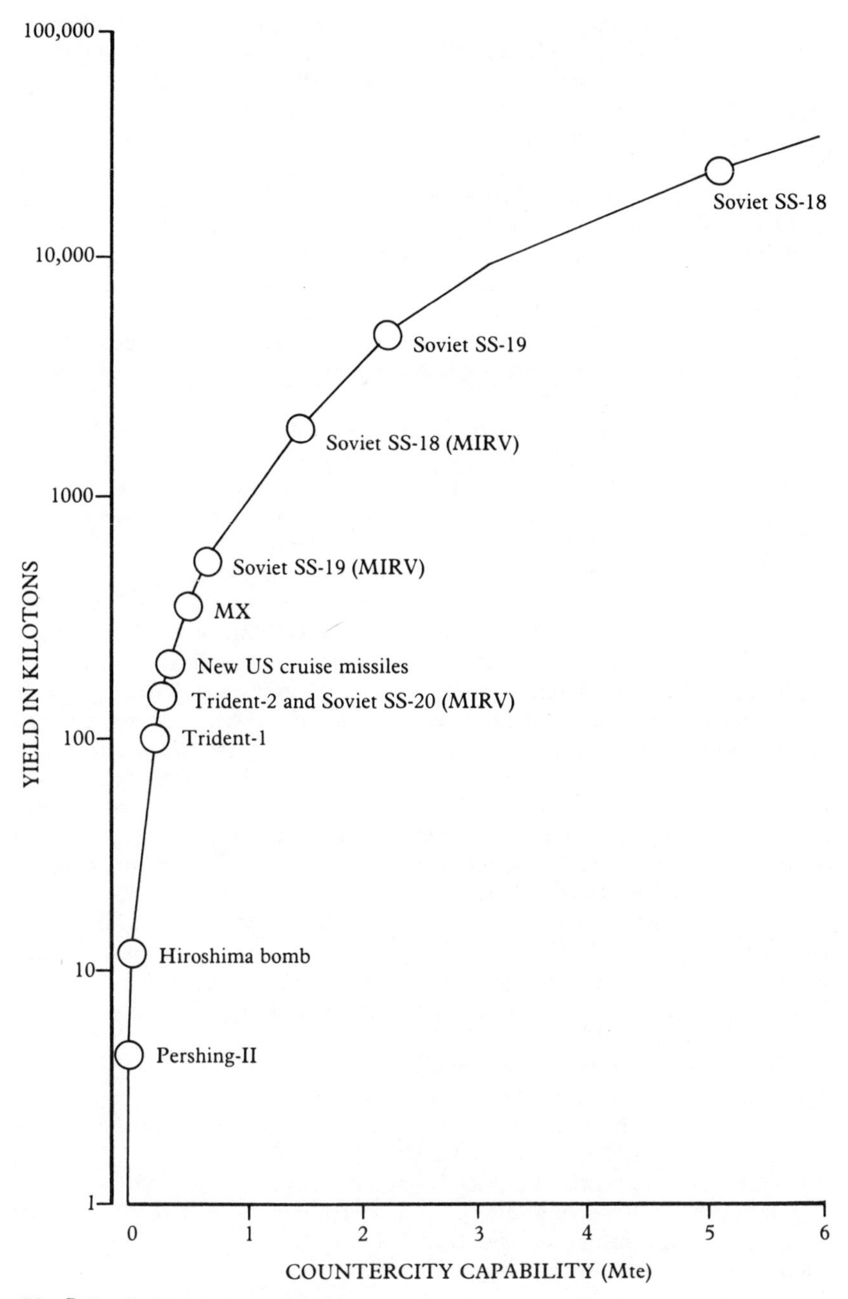

Fig. 7 Equivalent Megatonnage

The US Army's Lance missile is a tactical system designed for carrying either a 450-kilogram (1000-pound) conventional high-explosive warhead or a nuclear warhead of 10 kilotons weighing 212 kilograms (468 pounds). Lance's range is between 72 and 120 kilometres (45 and 75 miles) depending upon which type of warhead is carried, which makes the system 'tactical' in just about all practical circumstances; but a 10-kiloton yield is marginally less than that of the Hiroshima bomb, making 'tactical' more a matter of perspective as far as explosive force or prospective damage is concerned. Lance is launched from an amphibious tracked vehicle, the M752, which normally carries three missiles. Another mobile ground-to-ground-missile is the US Army's Pershing IA which has a range of up to 835 kilometres (520 miles) and carries a 400-kiloton warhead. Pershing IA is usually referred to as a tactical or theatre weapon but it is obvious that its range and yield make it a potential city killer. Its tactical classification refers more to targets such as airfields, for example, and other military sites rather than actual capability. Pershing IA will be replaced with Pershing II around the middle of the 1980s. The improved weapon will carry a warhead of between 1 and 10 kilotons but have a range possibly as great as 2415 kilometres (1500 miles).

Yield is not an accurate guide to differences between tactical, theatre and strategic nuclear weapons, save for those in excess of 500 kilotons. Weapons with this sort of yield are almost exclusively for use in strategic systems, but virtually all of America's strategic missiles carry warheads with yields far less than this. The context of use is a much clearer guide to classification than yield.

Although the combination of small yields and improved delivery systems have made the tactical or theatre use of nuclear and thermonuclear weapons more feasible from the point of limiting unwanted associated damage, a fundamental principle of modern military policy remains unaffected. This is the concept of the 'firebreak'; in other words, the crucial difference between nuclear and conventional chemical explosives irrespective of yield. Western doctrine emphasizes the need to postpone the use of nuclear weapons for as long as it is militarily practical to do so because, once the barrier has been crossed, the risk of escalation increases. Limiting any sort of war requires that both parties show deliberate restraint as a signal of intent. In this vein, the difference between nuclear and non-nuclear weapons is more discernible and less ambiguous than the difference between a small and a large nuclear yield.

Strategic Delivery Systems

Strategic weapons are currently deployed in three delivery systems, collectively known as the Triad: ballistic missiles launched from land sites, long-range bombers, and missile-carrying submarines. There are two essential criteria in assessing the appropriateness of each: penetrability and survivability. Penetrability requires that a significant proportion of the attacking force must be able to avoid destruction by enemy defences. Survivability requires that a significant percentage remain intact after being attacked. At the moment, fears for survivability underlie the most important developments taking place;

penetrability is less of an immediate concern, but is becoming more significant in the light of what appear to be promising technological developments.

Until recently, survivability has been guaranteed in three principal ways. Missile-carrying submarines hide in the vastness of the world's oceans, and, at the moment, it would be extremely difficult to locate and destroy a large percentage of those on patrol at any one time. Intercontinental ballistic missiles, the backbone of the land-based deterrent, are placed in especially hardened underground silos built to survive blast pressures of hundreds or possibly thousands of pounds per square inch. The survivability of strategic bombers is chiefly based upon quick reaction schedules aimed at getting them off the ground within a very few minutes of an attack warning and by keeping a percentage in the air at all times.

In 1977 President Carter took one of the most controversial decisions of his presidency by cancelling the B-1 bomber project designed to replace the Strategic Air Command's ageing B-52s. In 1981 President Reagan — as expected — reversed the decision and ordered 100 of an advanced version of the B-1 which would incorporate technical and design developments since 1977. The first squadron is scheduled for 1986. The first four B-1 prototypes built are supersonic aircraft capable of operating either at tree-top level or up to altitudes of over 18,300 metres (60,000 feet), reaching speeds of twice the speed of sound. In low-level approaches, which the majority of B-1 missions are intended to be, speed would be about 965 kilometres per hour (600 miles an hour) or just about the maximum speed of the B-52. Unlike the B-52, the B-1 is a variable-wing aircraft; its wings may be put into an extended or a swept-back position making it extremely versatile in speed and altitude. The plane is powered by four 13,600-kilogram (30,000-pound) General Electric two-shaft turbofans.

The B-1 prototypes are considerably smaller than their B-52 predecessors: a length of 45 metres (150 feet), height of 11.5 metres (38 feet) and a wing span varying with angle from just under 41.7 metres (137 feet) at 15 degrees to 23.7 metres (78 feet) at roughly 67 degrees, compared with 51 metres (167 feet) in length, 14.6 metres (48 feet) in height and with a wing span of 56.4 metres (185 feet). Depending upon type, an empty B-52 may weigh as little as 77,500 kilograms (171,000 pounds) or as much as 87,500 kilograms (193,000 pounds). The B-1 prototype, on the other hand, has an unladen weight of 63,500 kilograms (140,000 pounds). Both planes have a range of just over 9650 kilometres (6000 miles) with a maximum weapon load. But despite its smallness the B-1 has been designed to carry approximately twice the weapons load, which is also intended to be flexible for greatest variety: 34,000 kilograms (75,000 pounds) of free-fall bombs, 24 strategic air-launched thermonuclear missiles, or a combination of the two, carried internally. Externally, mounted on pylons under the wings, is room for 18,140 kilograms (40,000 pounds) of bombs or eight further missiles. These missiles would either be the short-range attack missiles (SCRAMS) currently deployed on the B-52 or the new AGM-86A air-launched cruise missile (ALCM).

The B-1's survivability is enhanced by construction designed to give it a greater resistance to nuclear blast effects and by incorporating features which minimize the time between alert and take-off. The plane is capable of standing at full readiness for an extensive period with regular and automatic checks. In an alert, the first of the four-man crew to reach the plane has only to throw a switch located behind the nose landing gear to fire its engines and activate all the systems necessary for take-off.

The plane has a highly sophisticated series of electronic countermeasure devices for jamming and confusing enemy radar. Something of the complexity involved can be seen in the ALQ-161 defensive system developed with the B-1 in mind although now being used to up-date the older and far less efficient defensive electronics on some B-52s. The B-1s tested with the ALQ-161 have an antenna located under each wing and in the tail, giving the system a 360 degree coverage. Each of the three antennae is linked to its own four different frequency band jammers with three transmitters each. Each sector is governed by a microcomputer, all three of which are linked to a central processing mini-computer which harmonizes and co-ordinates the activities of each in line with moment-to-moment assessment of the threats facing the aircraft.

In its flight, the plane is likely to encounter 100 or so different types of radar. Systems such as the ALQ-161 can assess each for its individual degree of threat and overall or relative importance at any given time in the flight. After this, the ALQ-161 will select a jamming response, the instant and span of the jamming and the power levels, frequency and modulation to be employed for each individual radar since it is capable of jamming several different types of incoming signal at the same time. It can interfere with one or more different fixed-frequency and variable-frequency radars simultaneously. Should it encounter a radar with new or unforeseen characteristics, the ALQ-161 will analyse it and select the most effective response. If it becomes overloaded with incoming radar signals, the system will drop threats with the lowest priority or reassign the respective jamming times assigned to them so as to be able to deal effectively with the most important. As the plane passes out of range of any radar, the system drops it from its 'threat-list' and adjusts to those of the moment.

The B-1 go-ahead is very much a stop-gap decision. Initially, the Strategic Air Command's requirement was for 241 aircraft at a total cost of some $77 million per plane. At today's prices, the cost is put at around $200 million per plane and will almost certainly escalate before the project is complete. Several hundred million dollars will be invested in developing a new Stealth bomber for deployment in the 1990s and the B-1 is aimed at modernizing America's strategic airforce in the intervening years. Stealth technology promises to make aircraft 'invisible' to radar by a combination of radically sharp aerodynamic contours and new coating materials which absorb or defuse incoming radar signals rather than reflecting them back. Stealth bombers, however, are likely to cost several times B-1's price and there is no guarantee that Soviet progress in radar and defensive technology will not make the whole project obsolete by the time it is perfected.

Considerations such as these make the future of the manned bomber the least certain of the Triad. Adherents of strategic aircraft emphasize the bomber's versatility — it can be launched, recalled or switched to alternative targets with ease. Unlike a ballistic missile, the manned bomber remains under political control for its entire mission. In the eyes of its critics, however, the bomber's chief weaknesses are a high degree of uncertainty that it would succeed in penetrating modern air defences or that a sufficient number would survive a first missile strike on its home bases. A missile strike from Soviet submarines could give American defenders a warning time as low as fifteen minutes. The reply is that reaction times are enough to get a high proportion of on-ground aircraft away and that the sophistication of modern penetration aids and electronic jamming equipment will insure that large numbers will reach their targets. The basic argument has been going on for several years and nothing — apart from an actual war — will settle the question of penetrability. All Reagan has succeeded in doing is to postpone the final commitment until Stealth's viability is known and to give the Airforce a partially modernized fleet in the meantime.

The Soviet Union has not placed any great emphasis on long-range strategic bombers since it perfected its first ICBMs. Russia's strategic bombers are roughly the same vintage as America's B-52s and lack the up-dating of advanced electronics. The Soviet fleet is made up of the Tu-20 Bear with a 12,550-kilometre (7800-mile) range, the Mya-4 Bison with a range of 9735 kilometres (6050 miles) and the medium-range Tu-26 Backfire with a range of 8850 kilometres (5500 miles). Backfire first entered service in 1974, making it the most recent addition to the fleet; Bear and Bison were first deployed in 1956 and Russia's other medium-range bomber, the Tu-16 Badger, in 1955. Backfire and Badger are essentially Euro-strategic systems but Backfire could be used against the United States if refuelled in flight or on a one-way suicide mission. This possibility made Backfire a stumbling block for a time in the second round of strategic arms limitation talks between America and the Soviet Union.

Although it is impossible to know for certain, it is highly doubtful if the Soviet fleet of strategic bombers, Backfire included, could penetrate American air defences to any significant extent. Soviet spending on strategic systems over the last decade or so has concentrated on the ballistic missile and this is likely to continue. Their long and medium-range bombers pose a genuine threat to Europe yet they would still be vulnerable to air defences.

Land-based missile forces are usually divided into intermediate-range to medium-range (IRBM/MRBM) and intercontinental-range (ICBM). Intermediate to medium-range missiles have a maximum range of some 6435 kilometres (4000 miles), anything above being classed as intercontinental. Missiles with ranges of around 800 kilometres (500 miles) or less are classed as short-range (SRBM). At present only France, the Soviet Union and China deploy land-based IRBMs/MRBMs. France's IRBM, the SSBS S-2, has a maximum range of just under 3220 kilometres (2000 miles). When it was first deployed in 1972 it carried a single 150-kiloton warhead which probably has not been significantly improved upon.

The Russians have three such ballistic missiles: the SS-4 Sandal, the SS-5 Skean and the SS-20. The SS-4 is now being replaced with the SS-20 and the SS-5 is practically obsolete. The SS-4 has a range of about 1930 kilometres (1200 miles) but the SS-20 can go some 2900 kilometres (1800 miles) or so further than that. The SS-20 carries either a single warhead in the megaton range or, in the latest model, three independently targetable warheads of 150 kilotons each. The latter is by far the most common. A four-warhead model with a yield of less than 550 kilotons each has recently been tested.

Many experts have argued that the SS-20 represents a sort of hybrid and is a technical violation of the 1972 SALT Treaty, SALT-I, which limited each side's numbers of strategic launchers. The first two stages of the SS-20 are in fact those of the SS-16 ICBM which is now deployed only in very small numbers if at all. To convert the SS-20 into an intercontinental weapon, supposedly all that is necessary is the addition of a suitable third stage in place of the one currently installed. America's cruise and Pershing missiles destined for Europe are the official answer to what seems to be a rapid build-up of SS-20s, although the final numbers deployed on either side of this Euro-strategic balance will depend upon the success or failure of efforts to negotiate another arms limitation treaty.

The United States, the Soviet Union and China currently possess ICBMs. China is very much the junior member, its missile still largely in the developmental stage. The CSS-X-4 is thought to have a range approaching 11,265 kilometres (7000 miles), which would enable it to target all parts of the Soviet Union and the western United States. It is said to carry a single 3-megaton warhead. By contrast, the United States and the Soviet Union currently deploy 1052 (normally 1054) and 1398 ICBMs respectively, numbers fixed in accordance with SALT-I. But there is little point in counting launchers without taking into account the number of warheads carried on them. This is why SALT-II was intended to fix the numbers of warheads on either side, but it will almost certainly never be agreed in anything approaching its original form.

ICBMs and all other strategic ballistic missiles come in three distinct types: those which carry a single warhead, those with multiple warheads (MRVs, multiple re-entry vehicles) and those classed as multiple independently targetable re-entry vehicles (MIRVs). An MRV package attached to a launcher effectively carries two or more warheads which separate in the final phases of flight to attack what is essentially the same target. MRVs can maximize damage by putting a number — often three — warheads into the same area, so creating what is called a nuclear 'footprint'. The maximum area which a single MRV package can cover is limited to a few thousand square kilometres, about 18,130 square kilometres or 7000 square miles in the case of the American and British Polaris submarine-launched missile.

With MIRVs, each of the warheads may be targeted separately. The MIRV package is often referred to as a 'bus' which is shot into space by the launch vehicle. At pre-programmed points in its trajectory, the bus releases its warheads at a speed, direction and time to direct it to its selected target. With the use of rocket steering, the bus is able to manoeuvre sufficiently to send its warheads over a far wider area than is possible with MRVs so that a single missile may attack targets throughout an area up to several tens of thousands of square kilometres. In effect, for MRV, the number of potential targets is equal to the number of missiles available, but with MIRV the number of targets can be equated with the total number of warheads carried by the missiles.

In 1980–81 the United States possessed three different ICBMs: Titan II, Minuteman II and Minuteman III. Titan II is by far the largest and the oldest, being first deployed in 1963. With a total weight at launch of some 150,000 kilograms (330,000 pounds), the missile stands nearly 31.4 metres (103 feet) high and has a diameter of 2.6 metres (102 inches). Titan carries a single warhead of 9 megatons. The 54 Titans — actually 53 after one blew up in its Arkansas silo in 1980 — are deployed in six squadrons of nine missiles each.

The two models of Minuteman are much smaller than Titan but newer; Minuteman II became operational in late 1966 and Minuteman III in 1970. Both models stand nearly 18.2 metres (60 feet) tall and have a first stage diameter of 1.8 metres (6 feet). Minuteman II has a launch weight of 31,803 kilograms (70,116 pounds) and Minuteman III 34,478 kilograms (76,015 pounds). Minuteman II carries a single warhead of between 1 and 2 megatons while Minuteman III carries a MIRV package normally composed of three independently targetable warheads yielding 340 kilotons each. This represents a recent up-grading of the system from a previous MIRV package of three 170-kiloton re-entry vehicles, and the final conversion to the new warhead, the MK12-A, was underway in 1982. In an overall programme of up-grading the missile, Minuteman III's accuracy was also improved by about 33 per cent, enabling any of its warheads to hit within a space of two football fields after an intercontinental journey. The United States deploys 550 Minuteman III and 450 Minuteman II.

To protect it from a Soviet attack each of the 1000 Minutemen and 54 Titans is housed in an especially hardened shelter or silo. The silos are designed to provide protection from a number of nuclear weapons effects including radiation and powerful electromagnetic waves but, in particular, blast. As of early 1981, Minuteman III was housed in silos constructed to withstand up to 1000 psi and Minuteman II and Titan II in 500 psi shelters. At the same time, there were reports of work being done on the 500 psi shelters aimed at bringing them up to Minuteman III standard. Silo emplacement not only gives the missile greater survivability but better accuracy as well since it enables it to be fired from a pre-surveyed site to a pre-surveyed target, so giving the guidance computer an exact set of co-ordinates. But its fixed location makes it vulnerable to increases in missile performance by the enemy because, if a sufficient combination of accuracy and yield can be obtained, the advantages of anti-blast hardening can be overcome. There have been an increasing number of reports that the Russians have obtained just such a capability, and American defence policy and procurement over the next decade will be devoted to overcoming what is seen as a vulnerability gap.

Soviet ICBM deployment in 1980–81 consisted of several types with several models each. The exact proportions of each vary from year to year as older missiles are replaced with newer ones and as those deployed are up-graded in accordance with improvements in warhead design and guidance mechanisms. In addition, estimates of yield and the numbers of MIRVed warheads vary far more for Soviet missiles than they do for America's. Four basic missiles compose the vast majority of the Soviet ICBM arsenal: the SS-11, SS-17, SS-18 and the SS-19. The oldest is the SS-11 Sego, which was first deployed in 1966 carrying a single warhead of between 1 and 2 megaton yield. By 1973, a second model had appeared carrying three MRV warheads of roughly 200 kilotons each. In 1980–81 a total of approximately 580 SS-11s of both models were deployed but the missile is steadily being phased out and should disappear by 1983.

The SS-11 has a maximum payload of about 680 kilograms (1500 pounds) which is comparable with the throw-weights of Minuteman II (450 kilograms or 1000 pounds) and Minuteman III (900 kilograms or 2000 pounds). The real throw-weight imbalance created by Russia's construction of massive and powerful launch vehicles is apparent with the remainder of the Soviet force. The SS-17 and SS-19 can carry maximum payloads of 2720 and 3175 kilograms (6000 and 7000 pounds) respectively and one SS-18 is capable of launching a warhead package weighing some 8165 kilograms (18,000 pounds) or more. Until recently, better American guidance technology and warhead design more than made up for the crude imbalance in payload but recent improvements in Soviet missile accuracy have eroded the difference to the point that both sides now have roughly the same capabilities and, according to the Pentagon, the advantage is moving to the Russians far quicker than anyone would have imagined 10 years ago.

As of 1980–81, the Soviet Union had deployed some 150 SS-17s, 308 SS-18s and 300 SS-19s. The remaining 60 ICBMs were a mixture of the older and obsolete SS-13 Savage and SS-9 Scarp and, according to some reports, a few of the mobile SS-16 emplaced in fixed silos. The majority of the SS-17s are assumed to be the Model I with four 900-kiloton MIRVed warheads but a few Model IIs with a single 5-megaton warhead may still make up a small proportion. The SS-19, which first appeared in 1975, has two main models: the Model I which carries a single 5-megaton warhead and the Model II which is MIRVed with a payload of six 550-kiloton warheads.

The SS-18 is the most controversial but little is known about it with any certainty. There are two main models. Model I carries a single warhead commonly estimated at between 18 and 25 megatons, the latest information putting it at 24 megatons precisely. Model II is the MIRVed version but estimates of warhead numbers and yields vary widely. The number of warheads is put at between eight and ten and their yields at between 550 kilotons and 2 megatons. Other versions have been tentatively identified. One has a single 20-megaton warhead but with an accuracy that shows only a slight improvement on the first model. Another is said to have a greatly improved accuracy (approaching that of

Minuteman III), the result of up-dating the MIRVed model and incorporating the latest Soviet research into missile guidance. A MIRVed warhead with at least 12 550-kiloton warheads has been constructed but not yet apparently installed because of the terms of SALT-II, which has so far been honoured despite not being signed.

The Soviets have developed cold-launch techniques for their latest ICBMs. Developed originally for firing missiles from submarines, cold-launching involves the use of a low-pressure gas to force the missile out of its silo (or submarine launch-tube) into the atmosphere where its first-stage propellants ignite. A steam injection method developed in the United States involves propelling the missile at some 130 kilometres per hour (80 miles an hour). At present, American ICBMs are fired directly from their silos where the intense heat and pressure cause extensive damage. Cold-launching reduces this damage to a minimum, which means that the silo can be 'reloaded' fairly quickly. It also eliminates the need for layers of protective shielding and vents to allow exhaust gases to escape. The extent to which the Soviets have deployed cold-launched ICBMs is not known for certain but the SS-17 at least is almost exclusively fired in this manner. The hardness of Russian silos is also widely disputed with estimates running from a few hundred psi to 1000 or 2000 in the case of large numbers of the latest SS-17, 18 and 19 models. A silo hardened to withstand 6000 psi overpressures is reportedly being developed for the next generation of ICBMs, due within a few years.

Although they have not yet been assigned numbers and model types, at least four new Soviet ICBMs under development have been tentatively identified by Western defence sources. Detail is still vague but one will have a payload similar to the SS-17, be highly accurate, based on solid fuel, be cold-launched and possibly intended for a super-hard shelter of 6000 psi. Another appears to be the size of SS-18 and carry at least the same payload. Both will almost certainly come in two models, one MIRVed and one with a single large-yield warhead. The larger missile will probably have a third version with three or four comparatively large-yield MIRVed warheads of 2 to 5 megatons each. The other MIRV may carry 10 or more highly accurate warheads of a yield between 400 and 900 kilotons. Other work in progress appears to be aimed at producing a missile approaching a height of at least 30.4 metres (100 feet), a diameter of 2.4 to 3 metres (8 to 10 feet) and the capability for a 4535-kilogram (10,000-pound) payload. This suggests a package of 10 or more warheads in the 550-kiloton class. Like the SS-20, this missile is probably intended for deployment on mobile launch-vehicles as well as silo emplacement.

The number of warheads carried by Soviet missiles very much depends upon the future of the SALT-II Treaty. Its original terms limit the number of warheads per launcher to 10. Although President Reagan has stated that it will not be signed in its present form, both sides have so far abided by the terms as set out. When this gentlemen's agreement runs out, the Soviet Union is likely to deploy MIRV packages of more than 10 warheads on its SS-18. According to some reports, such a package — carrying 13 or 14 warheads of about 550 kilotons each — has already been developed but, in the spirit of SALT-II, not yet deployed. With the possible exception of the new Trident submarine-launched missile, the United States will lack an immediate reply until the planned MX ICBM, currently under development, enters service in the late 1980s. If, as is possible, no new SALT treaty is ratified, MX may well carry more than the 10 340-kiloton warheads planned for it. Thirteen 300- to 340-kiloton warheads per MX would be a likely possibility.

Submarine-launched ballistic missiles (SLBMs) form the third and, for many, the most important part of the strategic Triad. At the present time, they are thought of as almost exclusively a deterrent force because the missiles lack the degree of accuracy associated with land-based missiles and the flexibility of manned bombers. This does not mean that they do not have potential military roles but only that their chief purpose is as a virtually invulnerable retaliatory force. Where the land-based ICBM's invulnerability is sought through emplacement in blast-resistant silos, the SLBM's security consists in the mobility of the nuclear-powered submarine patrolling beneath the world's oceans. Despite intensive efforts at perfecting means for locating and tracking submarines at sea, there does not yet appear to be any method by which the Soviet Union can do this effectively. Because of certain geographical and technological advantages, however, Western anti-sub techniques seem on the verge of being able to locate, track and destroy large percentages of Soviet subs at sea at any one time should the need arise.

The West's three principal SLBMs are Polaris, Poseidon and Trident I. Polaris is now deployed only by Great Britain, which maintains a fleet of four submarines carrying 16 missiles each. The last version of Polaris, the A-3, carries a warhead package of three 200-kiloton MRVed warheads and has a range of 4586 kilometres (2850 miles). In 1974 the British government began to investigate possible replacements for its Polaris force which was expected to become obsolete in the 1980s because of the age of the submarines and the likelihood of increasing difficulties in getting replacement missile parts as Polaris left active service with the US Navy. On the evidence available there seems to have been an early decision in principle to purchase the Trident system to replace the Polaris fleet in about 1990. At the same time, the decision was made to develop a new Polaris warhead to improve the missile's capabilities until Trident entered service with the Royal Navy. At a final cost of £1000 million, the new Chevaline warhead was scheduled to be fitted to Britain's Polaris missiles in 1981 but has been delayed by technical difficulties. In 1980 the government publicly announced its decision to purchase Trident at a cost officially put at around £5000 million but widely expected to be much greater.

The Chevaline package for Polaris contains six MRVed re-entry vehicles but how many of these contain live warheads is unknown. There could be three or four warheads of about 50 kilotons each and three dummies containing sophisticated electronic jamming and other devices designed to fool Soviet radar defences. The dummy warheads would also re-enter the atmosphere in a trajectory indistinguishable from the real thing. This would further confuse

Soviet defenders who would (hopefully) be unable to tell which of the six was actually harmless.

Although something of a mystery, the primary reasons behind the decision to invest considerable sums in a new warhead for Polaris seem to have been based on an original belief that the 1980s would see extensive Soviet deployments of anti-ballistic missiles which would drastically reduce the penetrability of the older Polaris system. The prevailing view is that such a small deterrent as Britain's must be able to threaten Moscow in order to have any pretence at independence. As a result, Chevaline was ordered as an answer to any major increase in the missile defences around the Soviet capital until Trident becomes available in the 1990s. A less charitable explanation, however, is that Chevaline is simply intended to give the Atomic Weapons Research Establishment at Aldermaston a project to occupy it for the decade before work on Trident can begin. Certainly, Chevaline does not amount to a technological breakthrough. Quite the contrary, in fact, for it is basically a refinement of an American system developed in the mid-1960s as an improvement for the Polaris warhead but which was discarded with the achievement of a MIRV system for Poseidon, America's second generation SLBM.

Poseidon entered service with the US Navy in 1971 as the first fully MIRVed SLBM. It carries 10 50-kiloton MIRVed warheads over a range roughly equivalent to Polaris and probably marginally greater. Poseidon weighs about 29,500 kilograms (65,000 pounds) at launch compared to 15,875 kilograms (35,000 pounds) for the Polaris A-3. It is 10.3 metres (34 feet) high and has a diameter of just over 1.8 metres (6 feet) compared to the Polaris A-3's height of 9.6 metres (31 feet 6 inches) and diameter of 1.3 metres (4 feet 6 inches). Poseidon may be carried by Polaris submarines after minor extension of the launch tubes and modifications have been made to the fire control system.

The Trident (C-5) I SLBM has the same dimensions as Poseidon but has a range of 7240 kilometres (4500 miles) and carries eight 100-kiloton MIRVed warheads. Trident's accuracy represents a considerable improvement on Poseidon and it is hoped finally to bring the warhead within 365 metres (400 yards) of its intended point of impact. Trident I is thus the first SLBM to have a genuine potential against some blast-resistant military targets as well as being a threat to cities. Some of America's Poseidon submarines are being refitted with Trident I in addition to the construction programme for Ohio-class submarines which have been specifically designed to carry the missile. The Ohio-class is the largest submarine ever to be built for a Western navy and weighs some 18,700 tons with a length of 170.6 metres (560 feet). This makes it longer than most of the US Navy's destroyers and some missile cruisers. The first of these ships — the USS *Ohio* — entered service in 1981 and President Reagan has commited the government to building one per year for the remainder of the decade. Each Ohio-class sub carries 24 missiles unlike the previous Polaris and Poseidon ships which carry 16.

In addition, Reagan has authorized the production of an improved version of Trident, the Trident (D-5) II, to enter service around 1990. Trident II has the same width as Trident I but is taller and cannot be deployed on anything but Ohio-class subs with especially designed missile launch tubes. Trident II can carry up to 14 150-kiloton MIRVed warheads and will have an intercontinental range of around 12,075 kilometres (7500 miles). This will enable America's sea-based deterrent to threaten Soviet targets from US coastal waters for the first time. Trident II will be the first SLBM to have a real hard-target killing ability and represents a clear decision in Washington to put an increasing strategic emphasis on the missile submarine. The key is the missile's accuracy; Trident II will be able to put any of its warheads within 182 metres (200 yards) of its target — roughly the capability of the improved Minuteman III ICBM. A likely future decision is to equip Trident II with Mk-500 Evader warheads capable of in-flight manoeuvring to avoid anti-missile defences and to home in on the target with an even greater accuracy. Known as manoeuvrable re-entry vehicles (MARVs), these systems are available for eventual deployment on America's full range of strategic ballistic missiles. Having commited itself to purchasing the Trident I, Britain faces the dilemma of incurring greater costs (£1000 million) in moving to Trident II or of deploying a system which the American Navy will be phasing out at the time it enters service.

The major problem with SLBM accuracy lies in the difficulty the submarine has in pin-pointing its speed and location to a degree sufficiently precise to compute a firing trajectory between the launch point and the known co-ordinates of the target. One proposed solution is NAVSTAR, a network of 18 to 24 navigation satellites which would enable the sub, or other naval vessel, to compute instantly its location anywhere in the world to within a few metres. The satellites, due to be placed in orbit by the American Space Shuttle towards the middle of the 1980s, may be geosynchronous. This will involve placing them at an altitude of some 37,100 kilometres (23,000 miles) where orbital speed will match the speed of the Earth's rotation and enable the satellites to remain over the same spot. An alternative is three rings of eight NAVSTAR satellites orbiting the Earth in periods of roughly 12 hours. This would allow four satellites to be drawn upon from any point on Earth. Each satellite will broadcast a finely timed signal. Upon receipt, the signal will be compared to precision clocks and geographical position will be computed by calculating the distance to each of the four satellites. Speed is computed through Doppler effects and the vessel's position relative to the satellites.

NAVSTAR will have a wide range of applications. Mobile, tactical missile crews will use the system to fix their location in the field and even individual troops may eventually carry portable NAVSTAR receivers. For strategic missile submarines, however, the system's importance lies simply in the greater degree of accuracy it will offer to SLBMs already deployed. Trident I, for example, would acquire an accuracy approaching that of currently deployed land-based missiles. In this way submarine-launched missiles — whose present accuracy is sufficient for targeting cities in a purely retaliatory role — will gain an increase in purely military capability by having far better chances of success against specifically blast-hardened strategic targets such as ICBMs emplaced in silos. It may also be possible to have signals from global positioning systems such as NAVSTAR interpreted by the missile's guidance mechanisms and on-board computer so that it can correct its course after launch and thereby gain greater accuracy.

The Soviet Union's principal SLBMs consist of the SS-N-8, which first entered service in 1972, and the SS-N-18, the most recent. It was the deployment of the SS-N-8, with a range of 7725 kilometres (4800 miles), that persuaded the United States to speed up Trident's development. The missile carries a single warhead of between one and two megatons but may be no more accurate than the now obsolete Polaris. The SS-N-18, on the other hand, has an accuracy which is capable of landing a warhead within 457 metres (500 yards) of its target and, with a yield similar to the SS-N-8, the missile has a potential military role that approximates Trident I's. The SS-N-18 carries three MIRVed warheads and has a range in excess of 8050 kilometres (5000 miles). According to London's International Institute of Strategic Studies, there were 160 SS-N-18s and 302 SS-N-8s deployed in 1980. The balance of the Soviet Union's 1980 SLBM deployment of approximately 960 missiles is composed of the older SS-N-5 and SS-N-6, which first entered service in 1964 and 1969 respectively, as well as a meagre 12 SS-N-17s.

At any given time, about 25 (60 per cent) of America's 41 strategic missile submarines are at sea. The equivalent figure for the Soviet Union is between 6 and 13 (10 to 20 per cent of its total of 64), and of these only a few are patrolling outside waters close to the Soviet coastline. In times of crisis, the United States may put upwards of 90 per cent of its missile subs on patrol while the Russians are thought to be unable to put more than a third to sea even in an emergency.

At the moment, America's anti-submarine technology is based upon a sound-surveillance system (SOSUS) which consists of a series of sound-detecting hydrophones placed on the sea bed. Soviet Yankee-class submarines carrying the older and shorter-range SS-N-5 and SS-N-6 missiles leaving their bases at Polyarniy on the Barents Sea or Petropavlovsk-Kamcvatskiy on the Pacific for deep-water patrols soon encounter specially laid SOSUS arrays which signal a detection to shore-based intelligence centres. Exit from the Barents Sea is effectively covered by two SOSUS networks running north from Norway to Bear Island and another stretching from Scotland to Greenland, with an intelligence centre at Scatsta in the Shetland Islands. A second SOSUS station in Britain is located at Brawdy in Wales to handle material from a detector array running into the mid-Atlantic. The Soviet Pacific base at Petropavlovsk-Kamcvatskiy is covered by SOSUS arrays running from the Aleutian Islands down through the Sea of Japan. Satellite links relay world-wide SOSUS data to the US Navy's anti-submarine intelligence centre in Norfolk, Virginia. The SOSUS system is said to be able to pin-point a submarine's location to within an area of about 64 kilometres (40 miles).

Several improvements to SOSUS are under development. One, a surveillance towed array system (SURTASS), uses surface vessels to patrol the oceans towing arrays of hydrophones. This not only gives the advantage of mobility but also extends the area of coverage since SOSUS is limited to the shallow waters of the continental shelf. Another development is sonobuoys dropped by aircraft, surface ship or submarines which drift to the sea bed and can be activated by signal. Work is also being done on long-range, continuous acoustic tracking systems for monitoring submarines in mid-ocean. Other research work is being carried out by America's Defense Advanced Research Projects Agency on the possibility of using satellites to detect and track submerged submarines by the heat patterns generated in their passage. Another potential area for submarine detection by satellite is through the distinctive wave patterns created in the ocean as the sub passes through. Otherwise, anti-submarine efficiency is being increased through more refined techniques of data processing and the integration of rapid satellite communications between sea-based detector equipment, local on-shore intelligence installations, central command centres and sub-killing aircraft, ships or hunter submarines.

Having long since recognized its growing disadvantage in deep water, the Soviet Union has concentrated upon deploying long-range SLBMs such as the SS-N-8 and the SS-N-18 which are capable of striking the United States from Delta and Typhoon-class subs on patrol in the Barents Sea or the Sea of Okhotsk. In both these areas the subs do not cross SOSUS barriers and so are relatively free from detection. This, however, is not likely to remain so for long. Either of the satellite tracking systems mentioned above would, if perfected, make submarines 'visible' even while on patrol in waters adjacent to the Soviet Union.

While the United States possesses a genuine strategic Triad of bombers, land-based ICBMs and ballistic missile submarines, it is doubtful if the same thing can be said for the Soviet Union. At present, any part of this American force is capable of surviving a Soviet attack in sufficient numbers to retaliate and devastate the USSR. Although enough Soviet bombers might survive an attack, it is unlikely that many would be able to penetrate American air defences. The Soviet fleet of strategic missile submarines is a real threat but is clearly inferior in capability and has been continually forced back into its home waters to ensure survivability in the event of war. When it becomes feasible for the US Navy to locate Soviet submarines patrolling in the Barents Sea or the Sea of Okhotsk and to destroy them, then the deterrent provided would at best be dubious. The principal Soviet strategic force therefore is mainly its land-based missiles and this probably goes a long way to explaining the massive effort that has been put into its steady development. This development may, as many claim, indicate a policy dedicated to obtaining an overwhelming strategic superiority but, on the other hand, it may simply reflect a realization that Soviet bombers and strategic missile submarines are now dangerously vulnerable.

w Strategic

MX: Controversy and Uncertainty

Just before leaving office, President Carter gave the long-awaited go-ahead for a new American ICBM. In late 1981 Reagan cut the initial numbers on order by one-half and rejected a controversial scheme to place the new missile on 'race tracks' spreading over thousands of square miles of American desert land. The source of this continuing debate and controversy is the most sophisticated missile yet, the 'MX' (missile experimental). Due to the improvement in miniaturization technology MX, with dimensions roughly mid-way between Minuteman and the ageing Titan, will carry between three and four times Minuteman III's payload and will contain far more sophisticated packages of guidance and jamming equipment, although weighing only 86,200 kilograms (190,000 pounds) and standing 21.3 metres high (70 feet). The payload will consist of 10 to 13 340-kiloton MK12-A MIRVed warheads. Ten is the current upper limit fixed by the un-signed SALT-II agreement but, if events create an unrestricted arms race, the final warhead package is likely to increase. The current plan is to deploy 100 MX by 1990 with 36 entering service in about 1986.

Any single MX warhead will possess a degree of accuracy previously unknown. It will be capable of hitting within 60 to 90 metres (200 to 300 feet) of its target after an intercontinental journey. This new-found accuracy derives from an Inertial Reference Sphere (AIRS) which consists of a 262-millimetre (10.3-inch) beryllium globe holding inertial guidance devices under precisely regulated buoyancy and temperature. These inertial guidance navigation devices are packages of gyroscopes and accelerometers which signal deviations in course and speed to the missile's computer which then commands the necessary corrections to the preprogrammed path. Technological refinements to inertial guidance systems will make MX twice as accurate as Minuteman III and will bring the capabilities of inertial guidance close to the theoretical minima imposed by the laws of physics.

However, there are other ways by which MX's guidance may be made even more precise. These are as yet undecided and depend on decisions about the basing system for the missile. One is the option of terminal guidance for each re-entry vehicle, which would give each re-entry vehicle a capability for navigational correction after scanning the ground below it, instead of following what amounts to a simple ballistic curve as happens at present.

MX is one of the most controversial defence projects in American history. There are two main reasons for this beyond its immense cost. The first has to do with the implications that follow from its greater accuracy, and these are detailed later. The second is the question of its basing, which was still in doubt in late 1981. The final result may well be an ICBM distributed among different basing systems. Equally, if problems or delays occur with the basing system finally chosen, one of the other candidates may re-emerge either as an alternative system or in a joint programme. The initial plan was to base 200 MX in complexes of 23 individual shelters each and randomly moving them in a nuclear 'shell-game'. An enemy, uncertain which shelter held the missile, would have to attack all 4800 to have a reasonable chance of destroying the missile force. At first, the scheme called for each shelter to be linked by a tunnel (the 'buried-trench' approach) but Carter altered it to a link of surface roads. The MX would have been carried by a large vehicle or, when the live missile was emplaced, dummy MX would have been transported emitting spoofing signals to fool Soviet satellite reconnaissance. The scheme was controversial from its beginnings; many believed it simply would not work and others — pointing out that it would involve the largest construction project in history — baulked at the price of at least $31,000 million plus yearly maintenance expenditures of $400 million.

Probably the greatest resistance came from the citizens of Nevada and Utah who were faced with the loss of an area of desert ranch land the size of Belgium. An MX complex would have to be extremely large because there must be a minimal distance between shelters. What this distance would be depends chiefly upon two things: the blast resistance selected for each shelter and some assumption on the yields of the warheads that might be involved in any attack. For example, at 500 psi (a very high estimate), each shelter would have to be separated by more than 970 metres (3180 feet) to ensure that a 550-kiloton warhead could not destroy two if aimed mid-way between. Against a 2-megaton yield the minimum distance would have to be nearly 1.6 kilometres (1 mile) and for 5 megatons more than 2 kilometres (1.3 miles). An attack by the gigantic 24-megaton SS-18 warhead would endanger two shelters unless they were more than 3.2 kilometres (2 miles) apart.

As a result of the land needed the Government faced continual delaying law suits that could have held up the project for years. The Pentagon considered endless alternatives; MX-carrying dirigibles were even suggested. But the alternatives to the multi-shelter approach came down to putting the missiles on refitted merchant vessels or on mini-submarines or aboard specially constructed aircraft. The submarine scheme was rejected for its cost, long development time and the fact that the US Air Force has little experience in operating subs and would be less than enthusiastic about handing its new missile over to Navy control. However, the mini-sub plan has been talked about for years and may reappear as an additional Trident II basing system. As conceived, the 350 to 1500 ton subs with a crew of 12 to 25 would have patrolled American coastal waters where they would be virtually invulnerable to Soviet anti-submarine efforts. Two or more cannistered MX (or Trident II) would be fitted to the outside of the hull and, when launched, rotated into a vertical position. Similar missile-carrying subs or small tracked vehicles operating in the Great Lakes have been suggested but would be unlikely to overcome environmental opposition.

The options are now basically three: emplacing them in super-hard silos, deploying them aboard special aircraft or defending silos with anti-ballistic missiles (ABMs). The dispersed-shelter system may also be reconsidered as it was favoured by the Air Force, and the government owns considerable land in the American South West where a reduced number of dispersed-shelter MX might eventually materialize. The possibility of ABMs is not restricted to any one possibility since they could be used in conjunction with any of the basing modes finally chosen — even to protecting airfields housing MX-carrying aircraft. ABMs, however, would require a serious re-negotiation of the terms of a 1972 treaty with the USSR which limits their deployment. The chief objection to hard-silo basing is that it was the supposed vulnerability of Minuteman silos that necessitated a radically new form of MX basing in the first place.

The first 36 MX will be housed in up-graded Titan silos and be cold-launched after the protective shielding is removed in order to accommodate the new missile. The Pentagon has not said what the new blast resistance will be but, from the language used, it will presumably be extensive — perhaps as much as the 6000 psi mentioned for new Soviet missile silos. The problem, however, is that there is a 'diminishing returns' effect for blast hardening as the graph on page 42 shows. If a Soviet 550-kiloton warhead had a 89 per cent chance of destroying a 1000 psi silo, for example, up-grading the silo to 6000 psi would reduce the chances of a Russian kill to 49 per cent. But, after going to the considerable trouble and expense of a 500 per cent increase in blast resistance, the Soviets need only improve their accuracy by 45 per cent in order to regain an 89 per cent likelihood of destroying with one warhead. For a long time there has been talk of deep-buried, hyper-hard silos where they would be generally invulnerable even to a direct hit and, apart from the initial 36, something on these lines seems to be under consideration for MX.

However, even if the silo could be constructed to withstand a direct hit — and this is by no means certain — there is the possibility that it would be so buried in rubble from the explosion, that the missile could not be launched. When the protective door slid away, for example, tons of rock and débris could fall in to block it. These problems exist at present and can supposedly be dealt with but future weapons might be aimed deliberately to engulf the silo in an impossible rain of radioactive earth and rock. There is also the question of future advancements in earth-penetrating techniques and the use of nuclear warheads to maximize seismatic shock. Having concluded that the fixed silo is now vulnerable to Soviet improvements in missile accuracy, the Pentagon is unlikely to be satisfied with the uncertainties of even vastly improved blast resistance without some other fall-back system. In this case, an eventual commitment to at least a limited deployment of ABM systems to protect MX is a distinct possibility. When one considers that the ABMs could find other uses — extending protection to divisions in the field, for example — the extra expense might appear justified. Anti-missile laser systems are another option but, as yet, the ultimate viability of the idea has yet to be proved.

Deploying MX on aircraft has the disadvantage of designing and developing

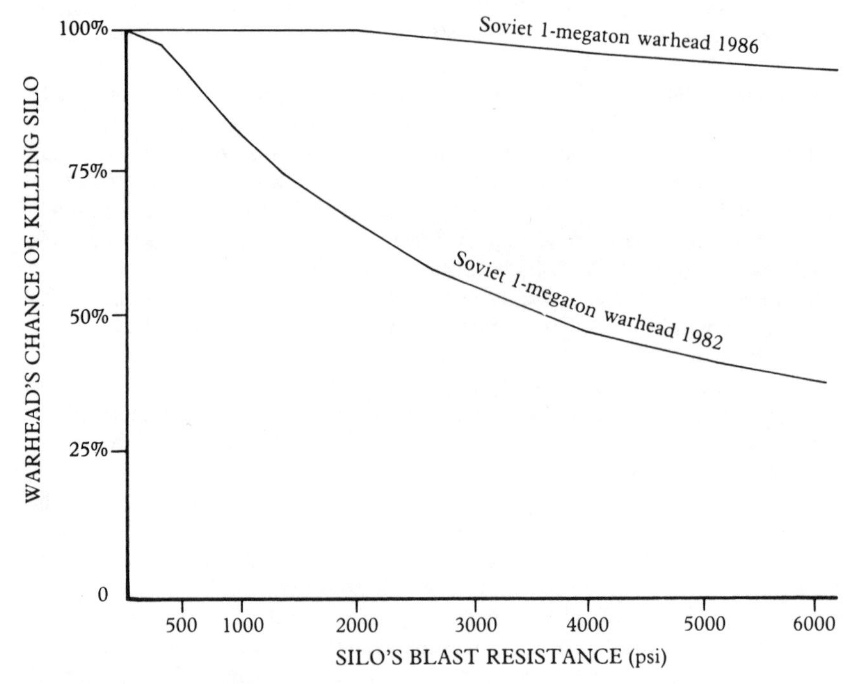

Fig. 8 Silo Vulnerability

Improving an existing Titan 500 psi silo to 5000 psi (+900 per cent) would increase its chances of surviving an attack by a 1-megaton Soviet warhead by only 35.5 per cent at current accuracies.

But, by the time MX enters service in 1986, if Soviet accuracies have been improved by 53 per cent — well within present-day capabilities — the silo's chances of survival are exactly as they were before.

a suitable aircraft — a project that would be both costly and lengthy — but, with a concerted effort, the planes could be ready at the time the missile comes on-stream. Aircraft deployed ICBMs have been considered for several years and successful tests with Minuteman carried out. For launching, the missile is ejected from the plane and ignites as it settles towards earth by parachute. MX aircraft would be designed for rapid take-off to escape fields under sudden attack and would be capable of remaining aloft for days at a time. There has been a great deal of criticism of the proposal on the grounds that MX's survivability would not be guaranteed but the system would be no more vulnerable than current B-52 and future B-1 squadrons which the Air Force steadfastly maintains are survivable. There would certainly be a secondary network of selected airfields — and even suitable stretches of Federal highway — where the planes could be dispersed in times of crisis. A portion of the fleet would always be aloft to ensure survivability as is presently done with strategic bombers and the numbers would be increased during crises.

The main objection to siting on aircraft is, however, the missile's accuracy. MX's precision is based upon its launch from a pre-surveyed site and dropping from a cruising aircraft would degrade its performance sufficiently to effectively negate what, in the Pentagon's eyes, is one of its most attractive features. It may be possible to get round this problem by placing a computer on board the launch aircraft to up-date its position continually via a land or space-based navigation system such as NAVSTAR. An identical feature incorporated into the missile's guidance computer would ensure both a reasonably precise fixing of the launch position and the ability to correct any inaccuracies during flight. But this would make MX's performance partly dependent upon an external system which might itself be attacked and neutralized. Once again, it does not seem likely that the Air Force will willingly accept a system vulnerability that would not be present in a land-based system. If some or all the 100 MX go aboard aircraft, terminal guidance (such as is discussed later for cruise and Pershing) is a likely addition as the warhead would not be dependent upon an external system for its accuracy. Improvements to internal guidance systems that would enable the launch plane to fix its position more accurately are quite possible but the current feeling is that achieving MX's full accuracy will require either land-basing, terminal guidance or external navigation corrections.

There are also political objections to aircraft deployment. Planes have been known to crash and the idea of having a few hundred thermonuclear warheads over the United States will not be popular. But patrols over the ocean will have repercussions in Europe where there is already an increasing resistance to the basing of new American theatre missiles. Fixed silos also have the advantage of being 'SALT-proof' in that the Soviets can be reasonably certain that the official number of ICBMs and the number actually deployed are the same. Even the dispersed-shelter system would have been verifiable as it was designed so the shelter covers could be periodically removed to allow Russian satellites to check that there was only one MX per complex. Aircraft numbers, on the other hand, are not so easily monitored. Without direct and reasonably frequent on-site inspection, cheating could not be discounted — especially in an atmosphere of international mistrust.

No final choice has effectively been made. Some commentators have even read this as a phase-down of the entire MX programme, and the cutting of the initial procurement by half and the obvious emphasis now placed on the Trident II programme support this view. With the installation of terminal guidance, Trident II would have the full counterforce capabilities expected for MX and the additional advantage of an acceptable and survivable basing system. The 100 MX could be emplaced in up-graded Titan and Minuteman II silos while the burden of maintaining a 'flexible' deterrent shifted to the Navy, but this would be resisted by the Airforce at all costs and is rather improbable. Of the various conflicting arguments over basing, ABM protection of siloed

MX is probably the most militarily attractive but the most politically dangerous. ABMs are an invitation to a technological arms race and highly controversial since they can be easily interpreted as an 'offensive' move to nullify the other side's deterrent rather than simply defending one's own. In the meantime, the final numbers of MX will await a final decision on their basing system — or systems — and may not be fully determined for several years.

The New Generation of Cruise Missiles

Both Hitler's *Wunderwaffen*, the V-1 'buzz bomb' and the V-2 ballistic missile, made deep impressions on the Allies during the last years of World War II, and as Germany was being overrun from both east and west, Soviet and American intelligence officers began an immediate programme of capturing all the abandoned equipment available and of recruiting scientists and technicians.

Of these two late inventions the V-1 was a small pilotless pulse jet aircraft with a range of almost 320 kilometres (200 miles). It carried 1 ton of high-explosive and was guided by an inertial unit of gyroscopes. Fired from northern France, it had an accuracy that was almost sufficient to guarantee hitting London, but the aircraft was vulnerable to anti-aircraft fire and fighter interception. The V-2 was a ballistic missile. Although it was more accurate than the V-1, and invulnerable to air defences, it carried a slightly smaller payload and cost nearly a hundred times as much per unit. After the war both the United States and the Soviet Union began to develop the principles of the V-1 into a family of nuclear delivery vehicles generally described as cruise missiles.

All cruise missiles are based on the same principle: small pilotless aircraft capable of navigating themselves by inertial guidance over several hundred kilometres to their targets. After the war the United States developed two main versions of strategic cruise missile, Snark and Matador, which were capable of being launched from ground bases or ships. About the same time, the Soviets began to deploy the first versions of their strategic Shaddock cruise missile which is still in service and carries a kiloton-range warhead some 725 kilometres (450 miles). It is principally deployed aboard submarines where it is designed to be fired from the deck.

Compared to advances in ballistic missile technology in the 1950s, the United States found cruise missile technology distinctly inferior and rapidly ran its programmes down. Then, in 1967, the use of a Soviet Styx shipboard cruise missile by the Egyptians to sink the Israeli destroyer *Elath* spurred renewed interest. Emphasis was first given to an anti-ship missile and the result was the McDonnell Douglas Harpoon with a range of up to 112 kilometres (70 miles) and a speed of almost 1050 kilometres an hour (650 miles per hour). Carrying a 225-kilogram (500-pound) high-explosive warhead, Harpoon may be launched from surface vessels, aircraft or, within a launch canister, from the torpedo tubes of submarines. After launch, the missile drops quickly down to cruise towards its target just above the water. In its final phases, it illuminates the target with radar signals and homes in on the image, climbing rapidly in the last seconds to dive onto the vessel. Harpoon is powered by a 300-kilogram

(660-pound) static thrust turbojet and is initially launched by a boost engine which accelerates it to near its full speed before shutting down and allowing the main engine to take over.

Work on the 4.5-metre (15-foot) air-launched version of Harpoon indicated the possibility of a new generation of long-range, highly accurate cruise missiles with nuclear warheads. At the same time the US Air Force was working on a flying decoy using cruise missile technology. It was known as SCAD (for subsonic cruise armed decoy) and was eventually cancelled, but, as with Harpoon, the work done showed the potential offered by the increasing miniaturization of guidance circuitry, warheads and jet engines. From the very beginning, the programme aimed at software integration and the production of a basic missile adaptable for both sea and air launching. Boeing Aerospace and General Dynamics competed for the air-launched version, General Dynamics entering a modification of its Tomahawk sea-launched cruise missile already selected for production. In May 1980 Boeing's model was chosen, General Dynamics having already won an additional prize with the acceptance of Tomahawk for deployment on the ground as well as on submarines and surface vessels.

There are now two main members of the new generation of cruise missile: the Boeing AGM-86-A air-launched weapon and the General Dynamics BGM-109 sea-launched and ground-launched versions of Tomahawk. Both will carry a 450-kilogram (1000-pound), 200-kiloton thermonuclear warhead and both will employ a sophisticated series of guidance techniques. Tomahawk will also have a tactical version of about the same weight and be equipped with a conventional warhead. Like Tomahawk, the Boeing missile's wings, tail assembly and air intake are enclosed within its body until about two seconds after launch when they become extended and the engine ignites to begin powered flight. Ground and sea launched Tomahawks are equipped with a rocket booster at the tail which fires the missile from its launch tube on the ship or ground vehicle until it reaches a height and speed to allow its own engine to take over, whereupon the spent booster drops away. Tomahawks fired from submarines are enclosed in a canister which is fired through the torpedo tubes and up to the surface where the booster ignites, forcing the missile out of the canister and to an altitude where it can begin powered flight.

Both the Boeing air-launched cruise missile (ALCM) and the Tomahawk have subsonic speeds of between 800 and 960 kilometres an hour (500 and 600 miles an hour) and employ a 270-kilogram (600-pound) thrust turbofan jet engine weighing about 65 kilograms (144 pounds) and measuring only 0.9 metres long by 0.3 metres wide (3 feet by 1 foot). Tomahawk is the larger of the two missiles with a length of 6.4 metres (21 feet) including the booster and a diameter of 530 millimetres (21 inches). It has a maximum range of around 120 kilometres (760 miles) but, with the addition of an extra fuel tank mounted on the belly, this can be extended to about 1600 kilometres (1000 miles). With the 225-kilogram (500-pound) tank attached, its launch weight is 1088 kilograms (2400 pounds). Tomahawk's launch weight varies with the mode of deployment, the ground-launched version weighing about 1360 kilograms

(3000 pounds), the submarine version 1815 kilograms (4000 pounds) and the air-launched tactical version around 1130 kilograms (2500 pounds). Tomahawk's range is in excess of 2410 kilometres (1500 miles) but the tactical model has a maximum of only 560 kilometres (350 miles). The tactical model is intended largely as an anti-ship missile but will be useful for unmanned bombing missions and aerial surveillance as will the strategic version when the thermonuclear warhead is replaced with observation and reconnaissance equipment.

The US Navy tested a robotic bomber variant of Tomahawk in 1978. The missile flew 650 kilometres (403 miles) to its assigned target, an airfield at a Utah test range, and dropped 11 of 12 simulated bomblets on a centre line target on the runway. It then flew back to simulate a damage-assessment overflight.

The Boeing ALCM and each of the Tomahawk variants will employ at least two advanced navigation techniques and possibly three. The first is an improved package of accelerometers and gyroscopes for inertial navigation. The second is a system known as TERCOM (for terrain contour matching). The third possibility would be satellite positioning via NAVSTAR. In addition to these, both missiles will have a terminal guidance technique known as SMAC (scene matching area correlation). All these will make Tomahawk and the ALCM the most accurate strategic missiles ever developed so that after a maximum journey they will be capable of impacting within an area of incredibly no more than 6 to 12 metres (20 to 40 feet) from their intended target. At their best, this makes them two and a half times as accurate as MX.

Although the inertial and TERCOM devices are more than sufficient to steer the missile to its target, NAVSTAR would nevertheless provide a more precise means of position fixing than the inertial unit alone and could fill in should anything go wrong within the missile's internal systems. Because its satellites can be destroyed or their signals jammed, NAVSTAR could never fully replace self-contained navigation systems within the guidance package built into Tomahawk or the ALCM but it would be able to supplement it with exceptional precision. By drawing upon the satellites, the missile will be able to determine its exact position at any time during its flight with a margin of error of no more than 9 metres (30 feet). Otherwise, to improve on its inertial guidance, the missile will make corrections based solely upon TERCOM.

Both Tomahawk and the Boeing ALCM will contain a highly sophisticated radar altimeter located in the nose. During flight the altimeter will constantly bounce a range of signals off the ground, enabling the missile to engage in a ground hugging flight pattern but still vary its altitude in line with the terrain below. Over flat ground the missile will fly at an altitude of about 15 metres (50 feet), climbing to 45 metres (150 feet) in hilly country and between 97 and 105 metres (320 and 350 feet) in mountains. By constantly adjusting its height to the terrain, the missile is able to follow a tree-top path and 'hide' itself in the ground clutter that renders most anti-aircraft radars ineffectual at low altitudes. Thus, even though the missiles fly at fairly low speeds, they are difficult to spot and intercept. In addition, they have been designed to give minimal radar profiles, as aerodynamic structures reduce signal reflections to a minimum. New computer-integrated navigation systems will programme cruise to fly through radar 'blind spots' created by prevailing weather conditions. Course and altitude updates will be available up to launch. 'Stealth' coatings with materials designed to absorb or diffuse illuminating radar impulses rather than send them back in coherent patterns are a possible refinement.

The radar altimeter will also correct the unavoidable navigational inaccuracies introduced by inertial guidance. Built into the missile's memory banks are 20 to 25 terrain profiles converted from satellite photos into grid maps with each grid assigned a number representing the altitude of that sector. To up-grade its inertial navigation, the missile takes a radar reading of the ground below it and grids it into a digital map of altitudes. It then correlates that reading with the information stored in its memory, determines exactly where it is and corrects its course as required. At least four of these TERCOM up-dates would be used for any given flight, but the extra TERCOM data in the missile's memory will be used for alternative course provision and for programming a random-path flight — or 'zig-zag' approach — to make it as difficult as possible for an enemy to make an interception by predicting course and target.

As it reaches the target area, guidance will be assumed by the terminal correlator. As in the TERCOM system, the on-board computer contains a reference grid but in this case it is of the target itself. By scanning the area with a variety of radar, optical and infra-red sensors, the missile will pin-point its exact objective by comparing the data with the information in its memory and then fly unerringly towards it. The system most commonly spoken of for Tomahawk and the ALCM is DIGISMAC, which compares a digital conversion of the satellite photo with a similar conversion of incoming optical data. It is likely to utilize two or three frequencies covering both the infra-red and visible parts of the spectrum. Modern radars are capable of distinguishing objects as little as 0.3 metres (1 foot) from their backgrounds and building up a profile for explicit identification.

Prior to launch a Tomahawk or ALCM will receive minute-by-minute updates on the exact position of the launch platform, whether on an aircraft, submarine, surface ship or ground vehicle. The on-board computer thus knows where the missile is at any given time and is ready for a virtually instantaneous firing. Over the ocean, very flat terrain or snow-covered ground Tomahawk and the ALCM will probably have to rely exclusively on inertial guidance but NAVSTAR would extend the types of territory it could traverse with precise navigation. Alternatively, Tomahawk might use a system which would exploit variations in the Earth's magnetic field, known as MAGCOM (magnetic contour matching), but this system is not yet fully developed and any initial improvement on the inertial-TERCOM package for non-distinct terrain and over water will be through satellite navigation.

By 1980 several spectacular failures in the testing of Tomahawk led some writers to jump to the conclusion that the new missiles would not work. Certainly, deployment scheduling suffered setbacks and the 1983 goal for stationing the ground-launched Tomahawk in Europe will not be met. TERCOM was criticized because it involved comparing optical information obtained from a satellite photo with a radar profile of the ground — in other words two distinctly different things were being compared rather than like with like. However, it is not a question of establishing an exact identity between the maps encoded in the missile's memory and the terrain below. Instead the computer establishes a high correlation between the two and infers its location with a virtually certain probability. Apart from flat terrain, the real difficulties come from dense foliage and snow which may effectively fool the computer when it computes its altitude and location. Seasonal and geographical weaknesses in TERCOM's ability to compute its location can probably be satisfactorily solved by an improved inertial guidance, MAGCOM or enabling the missile to receive NAVSTAR signals when the computer is in doubt. Altitude may be a more serious problem since cruise missiles are designed to fly as low as possible and if such things as snow continue to cause false readings, the missile's effectiveness will be reduced dramatically.

Due to the setbacks in development scheduling, the 464 ground-launched Tomahawks planned for Europe are unlikely to be in service until 1985–86. Britain has already agreed to take 160 ground-launched cruise missiles (GLCMs) for basing at Greenham Common and Molesworth Air Force bases in the south-east of England. The remainder are destined for West Germany, Holland, Belgium and Italy but, in late 1981, these countries had yet firmly to agree to allow Tomahawks to be based inside their boundaries. The missiles would be under exclusive American control and deployed on manned, mobile transporter-launch vehicles (TELs) with four Tomahawks each. After attack warnings or during periods of acute crisis, the TELs would disperse throughout the countryside, pulling off the road in isolated locations and awaiting orders. With sufficient warning, the British-based missiles would spread over an area stretching from the Midlands to Cornwall.

The GLCM is a theatre or Euro-strategic weapon only because the mobility of the launch vehicle is restricted. The shorter range, anti-ship tactical version of Tomahawk was scheduled for deployment aboard US Navy submarines and surface vessels in 1982–83 with the long-range, thermonuclear-armed version following roughly a year and a half later. At the moment, America's nuclear subs can take seven to ten sea-launched Tomahawks but could be modified to take between two and three times that number. The numbers deployable on surface vessels varies from around 50 for a destroyer to 100 or so for the larger cruisers and aircraft carriers. Adopting merchant vessels as clandestine carriers of Tomahawk would be simplicity itself — one of the main reasons the missiles have been described as the arms treaty cheat's dream weapon.

The Boeing air-launched version is destined for deployment on the Air Force's B-52 and FB-111A strategic bombers. The FB-111A is a two-seat, supersonic bomber with a range of 5090 kilometres (3165 miles) extendable with the addition of external fuel tanks. It will carry up to two ALCMs within its bomb bay and four mounted on pylons under the wings. The B-52, however, will hold up to eight fixed upon a rotary launcher in the aft bomb bay and as many as 12 on wing pylons. The B-52's forward bomb bay holds four thermonuclear gravity bombs. The rotary launcher was originally designed for the Boeing SCRAM (short-range attack missile) and works on the revolver principle with the motor turning the launcher to bring each missile into firing position. The launcher lines up each missile with the doors of the bomb bay, then, on firing, drops it from the aircraft where it ignites and follows its pre-programmed course to the target.

SCRAM is a 4.2-metre (14-foot), 1010-kilogram (2230-pound) wingless rocket which travels about three times the speed of sound. It carries a 200-kiloton thermonuclear warhead and can be used to attack targets from about 55 kilometres (35 miles) to slightly over 160 kilometres (100 miles) away. Although guided internally it also contains a terrain avoidance radar enabling it to follow the contours of the ground in a low-level approach, and it can also be launched in a ballistic trajectory to its target. A B-52, or future B-1, attack approach on the Soviet Union would involve combinations of SCRAMs, gravity bombs and ALCMs. As the aircraft approaches within range of its assigned targets it would begin to launch ALCMs at targets between 160 and 1225 kilometres away (100 and 760 miles distant), SCRAMs at those within 160 kilometres (100 miles) and drop gravity bombs on targets within the actual flight pattern. Targets up to 1600 kilometres (1000 miles) away could be attacked with ALCMs equipped with an auxiliary fuel tank but these missiles would not fit on the rotary launcher.

Currently under development, the ASALM (advanced strategic air-launched missile) includes most of the features of SCRAM and the ALCM. Its speed is likely to be around four times that of sound and it will have a range four or five times greater than SCRAM. Like SCRAM, its guidance will be inertial but a TERCOM facility will be added. When developed it will be both a strategic version carrying a thermonuclear warhead and a tactical model equipped with a conventional explosive. As a strategic weapon, the missile would be launched against ground targets but its tactical version would be aimed at high value aircraft such as airborne early warning radar systems. ASALM would be propelled by a combination of integral rocket and ramjet. Upon ignition, a solid fuel rocket would accelerate the missile to flight speed and, on burning out, would jettison the exhaust nozzle and air-intake covers turning the engine unit into a ramjet and the ASALM into a cruise system.

An advanced version of the Boeing ALCM is also under development. It is aimed at bringing the operating range (without extra fuel tanks) into line with Tomahawk's and, on increasing its speed, possibly into the supersonic category. This new ALCM-B would be 1.5 metres (5 feet) longer (Tomahawk size) and therefore incompatible with the current B-52 rotary launcher and the FB-111A bomb rack. B-52s would therefore carry ALCM-Bs only on wing pylons but a modified B-1 will probably be equipped with a new external launcher which could accommodate SCRAM, the ALCM and the ALCM-B.

Many people still associate a bomber with an attack on one or two cities, but a B-52 could attack as many as 24 cities with a full package of SCRAMs, ALCMs and gravity bombs. In practice two or three weapons would be assigned to the same target but that would still involve the devastation of eight to twelve cities by a single aircraft. The first design for the B-1 called for twice this damage potential and current modifications will probably increase it. In the War years the Luftwaffe could buy about 300 V-1s for the price of a single British Lancaster bomber. Today the price gap is narrowing as Tomahawk and the ALCM will cost roughly $1.5 million each compared to a unit price for the B-1 of $200 million. Now that a new bomber appears certain, air-launched cruise missiles are expected to increase the flexibility and destructive power of the next generation of aircraft beyond anything the ALCM's designers first envisaged, but only so long as they work as originally predicted.

Equally, the US Navy's experiments indicate a real chance of robot bombers attacking airfields or battlefield targets. They could also be employed as robot drones surveying a battlefield with a variety of optical and electromagnetic sensor equipment to designate targets for air strikes and artillery or simply to send back vital intelligence without risking a costly aircraft and pilot. But their immediate role is that of a strategic or theatre nuclear delivery vehicle. With the performance foreseen, cruise missiles could attack any target effectively and they should have the accuracy to destroy the hardest military construction as well as the largest city.

Pershing II: The Strategic Question Mark

All the controversy surrounding the Soviet SS-20 IRBM and the American package of Tomahawks destined for Europe has led to neglect of the 108 Pershing IIs which are scheduled to be deployed in West Germany before the middle of the 1980s. In fact, in several ways Pershing is a far more disturbing possibility than Tomahawk but it is the cruise missile that has attracted all the publicity. The attitude is understandable since Pershing is a ballistic missile and has been fielded since 1962 while Tomahawk's terrain reading capabilities make it seem more of a technological breakthrough. But since Pershing was first deployed, there has been a steady programme of improvements which will culminate in the Model II, due to enter production in 1983.

The re-entry vehicle developed for the Pershing II will have terminal guidance and accuracy error reduced to between 12 and 36 metres (40 and 120 feet). Using the Correlatron developed by Goodyear Aerospace, the re-entry vehicle's memory banks will contain a radar profile of the target gathered previously via satellite or aerial reconnaissance. Within the ceramic nose cone a narrow beam radar operates when the re-entry vehicle reaches about 15,240 metres (50,000 feet); an antenna begins to turn at two revolutions a second, reading the area below in a circular pattern. Having been brought over the target area by the inertial guidance unit, the Correlatron examines the ground below, locates the target by comparing its readings with the data in the memory banks and signals course corrections — via the inertial guidance unit — to hydraulic steering fins located in the re-entry vehicle's tail. Differences between the two profiles are detected by an electronic multiplier and the two are correlated to locate the target from the surrounding area. Although defence systems can jam out some incoming signals the designers claim that they cannot eliminate a sufficient number of them to make the Correlatron ineffective.

Whether the Pershing re-entry vehicle will be fully manoeuvrable is not yet known for sure. Manoeuvrable re-entry systems (MARVs) have been on American drawing boards for years but have not yet been produced, largely because of treaty limitations on anti-missile systems. If Pershing is equipped with one it will be capable of full powered flight in its terminal phases rather than only being able to adjust its ballistic trajectory. The missile would be launched on a ballistic trajectory aimed *away* from the impact point, but the guidance unit would be programmed to locate the target via the Correlatron during re-entry and to initiate a terminal course change in the final moments before detonation. A MARV will contain radar for detecting an attack by a defensive missile and be capable of taking evasive action while constantly homing in upon the target.

Because of its immense accuracy, Pershing II can be equipped with extremely low-yield warheads yet still fulfil any desired mission; unlike less accurate weapons, it will not require large yields to make certain that a target is destroyed in the event of a wide miss. As a result its warhead strengths are estimated at between only 1 and 10 kilotons, giving it a tactical yield, but its role will be strategic on account of its range and accuracy.

The full implications of Pershing II's accuracy are detailed later but, basically, the missile can destroy anything it is fired at — even if the target has been specially constructed to withstand nuclear weapons. Furthermore, Pershing's range appears to have been extended by as much as 2415 kilometres (1500 miles), putting targets in western Russia, even Moscow, under threat from missiles based in western Europe. 1980 estimates gave Pershing II a maximum of between 1600 and 3200 kilometres (1000 and 2000 miles) with something around 2415 kilometres (1500 miles) being the most commonly quoted.

Soviet strategic weapons targeted upon western Europe consist of the SS-4 Sandal, which carries a single 1-megaton warhead over a maximum range of 1930 kilometres (1200 miles); the SS-5 Skean, with a single warhead of the same yield but a maximum range of 3700 kilometres (2300 miles); and the SS-20, with three 150-kiloton MIRVed warheads and an estimated maximum range of from 4025 to 5630 kilometres (2500 to 3500 miles).

The furore that followed upon the SS-20's initial deployment centred largely upon its MIRV capability, as a one-for-one replacement of the older 600 single-warhead SS-4 and SS-5 launchers would mean a total of 1800 warheads targeted upon NATO. However, as yet there is no evidence that Moscow intends to go this far. Some suggest that the numbers of SS-20s put into service will stop at around 200 to maintain the same number of warheads when the older launchers are fully retired. Others disagree and suspect the Soviets of attempting to tip the European warhead balance heavily in their favour. The ground-launched Tomahawk and, to a lesser extent, the improved Pershing have been put forward as a counter to the SS-20 and as 'negotiable' in the sense that numbers to be finally deployed will depend on future agreement regarding the SS-20.

Pershing II, however, is a qualitatively different weapon from the SS-20. Although the SS-20 has a yield and accuracy (launched from a pre-surveyed site) that enables it to attack a wide variety of civilian and military targets successfully, used in Europe it would be targeted upon airfields, naval bases and population centres and it does not have the range to attack the vitally important ICBM complexes in the American Midwest. Pershing II, on the other hand, will have the range from West Germany to reach some Soviet ICBMs targeted upon America and the accuracy to destroy them with virtual certainty. This makes the planned Pershing a potentially strategic weapon of much greater importance to Washington than merely as a contribution to the balance of power in Europe.

4 New Capabilities and the Calculus of Limited Nuclear War

Missile Targeting

Strategic doctrine now posits the possibility of nuclear warfare which would involve essentially military targets only. The key terms are 'counterforce', which means attacking solely military targets, and 'countervalue', the deliberate destruction of a nation's population and industrial capability, and strategists argue whether or not the first is realistically possible without inevitably leading to the second. In effect, nuclear war is now looked upon in terms of a spectrum of severity which runs from 'limited' strikes upon purely military targets to an all-out city-annihilating catastrophe.

The name for this doctrine has changed from Defense Secretary Robert McNamara's 'Graduated Response' in the early 1960s to 'Flexible or Selective Options' in James Schlesinger's term of office during the Nixon presidency. Although the basic concept has altered little, the military capabilities have. The one essential key is the increased accuracy of ballistic missiles which has not only made counterforce plausible but, in doing so, has indirectly altered the assumptions upon which just about all the world's major arms agreements and security policies have been based for the last 20 years. Basically, what increased accuracy does is to increase the range of possible targets and extend options into a specifically military sphere that less accurate weaponry would not be able to damage with any reasonable certainty of success.

The characteristics of nuclear explosions create two essentially different types of target, 'hard' and 'soft'. The distinction is generally the amount of blast overpressure (in psi) that it takes to destroy them. While there is no absolute line dividing the two, 'hard' can be thought of as beginning somewhere in the 20 to 50 psi range and running upwards into the thousands. A distinction with much of the same meaning is that between 'area' and 'point'. The latter means literally what it says; it is a mathematical 'point' which is aimed at for a direct hit. 'Area', as the name suggests, puts far fewer demands upon accurate guidance. Again there is no absolute distinction but 'hard' and 'point' are generally associated and are contrasted with 'soft' and 'area'.

A city is an area target because it may cover several hundred square kilometres. Putting a warhead somewhere within that space is not particularly difficult, even if it misses the intended ground-zero. It is also a soft target because the blast, heat and radiation of even one nuclear explosion will cause appalling devastation to life and property. Any number of buildings within it may have a sufficient blast resistance to survive virtually intact but the intended 'target' is not any particular construction but the city itself. Missile silos and buried command bunkers, however, may easily survive if the warhead impacts more than 60 or 90 metres (200 or 300 feet) away.

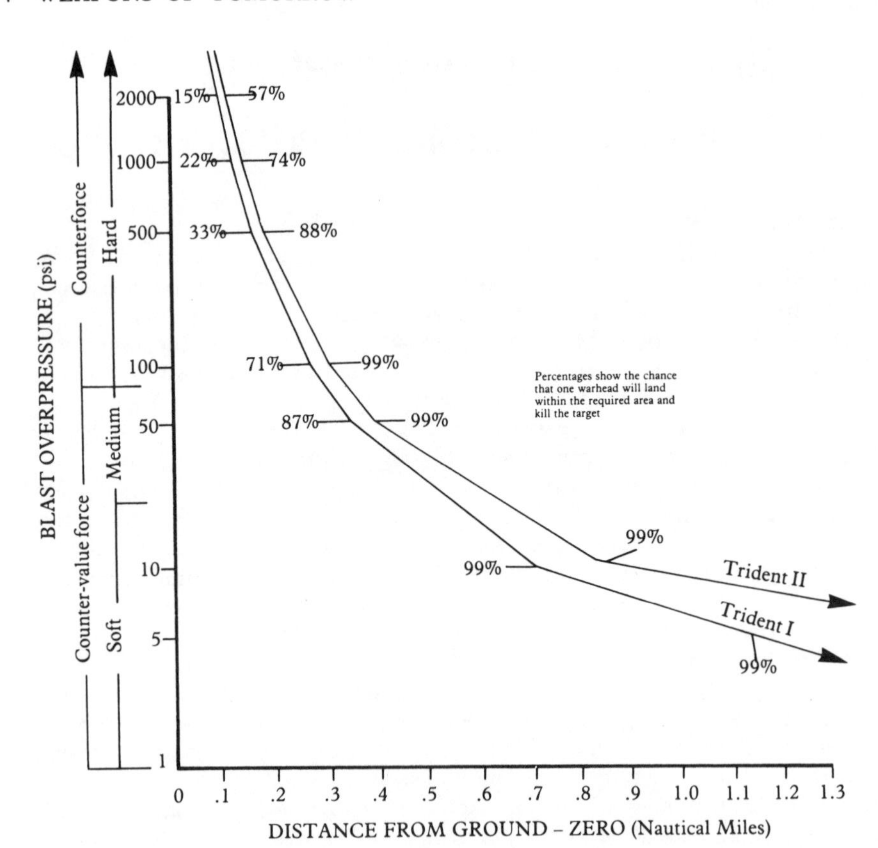

Fig. 9 Tridents I and II Targeting Capabilities

Counterforce is not exclusively involved with hard targets; nor is countervalue always a question of soft or even area sites. There are many types of essentially non-military targets which are extremely resistant to blast — air raid shelters, hydroelectric dams and industrial machinery, for example. Equally, many purely military targets such as radar antennae, unsheltered aircraft and massed infantry are highly vulnerable to blast, heat and radiation. Assuming 7 psi to be sufficient, a 1-megaton warhead could miss its target by over 3 kilometres and still destroy an open aircraft. To destroy a blast-resistant construction hardened to withstand 600 psi, on the other hand, the warhead must come within about 550 metres (600 yards).

The world's missiles are accurate enough for most countervalue targeting, but there are exceptions. Important industrial complexes, for example, require precision guidance, as would buried command centres housing a country's leaders. In general, however, 'countervalue' refers to population, socioeconomic institutions and industrial potential which, with the exception of agriculture, are concentrated in urban areas.

The essence of countervalue in deterrence lies in being able to inflict 'unacceptable damage' on the enemy — a level of damage to his own country which no rational political leader would run the risk of incurring in retaliation to an offensive he might launch himself, for whatever reason. Former US Secretary of Defense McNamara once outlined what America estimated to be unacceptable damage for the Soviet Union as the death of between 20 and 25 per cent of her population and the destruction of one half to two thirds of her industrial potential. The ability to inflict this level of damage remains the basic rule of thumb for determining America's second strike capability — the proportion of the strategic Triad which would realistically be expected both to survive any initial attack and succeed in then penetrating Soviet defences.

For China, McNamara's criterion was an ability to kill half the urban population. The difference centred largely on a belief that the value of China's relatively small industrial and professional workforce was of far greater importance to her than crude numbers would indicate. Equally, certain cities in all countries have a symbolic value greater than simple economics indicate. This possibly gives a small deterrent force such as Britain's more effectiveness than might be supposed.

The graph on page 54 shows that concentrations of Soviet power protected by blast resistances of 300 and above would be reasonably safe from Trident I but not from the Trident II which Britain will almost certainly buy instead in order to be able to strike at Soviet Silos. The Soviet political system itself has always been considered 'targetable' in Western plans on account of its high centralization, and priority targets include the administrative and Party centres which form direct links with the central authority in Moscow.

City Targeting

'Soft' urban targets involve uncertainties. Predicting the exact damage in a given city on a given day is more difficult than often thought because of the many different variables. Climate, terrain and prevailing types of urban construction are just a few. The effects of a bomb's heat, for example, drop dramatically on overcast days. Equally, the time of the attack may be critical since more people are at risk at mid-day during the working week than they are at the weekend or on public holidays. Because of these uncertainties, generalized death forecasts are usually based upon blast effects which are more predictable than heat or radiation damage in different circumstances.

A simplistic but nevertheless useful approach to forecasting urban fatalities is to use the Rand Bomb Damage Computer Model which was originally developed from the results of Hiroshima and Nagasaki. The model predicts blast (immediate) fatalities only and leaves out those due to thermal and radioactive effects. By doing so, it understates the actual number of deaths by a factor of two to three if not more in some cases. It also gives no idea of levels of casualties, social disintegration and general misery that would follow even the

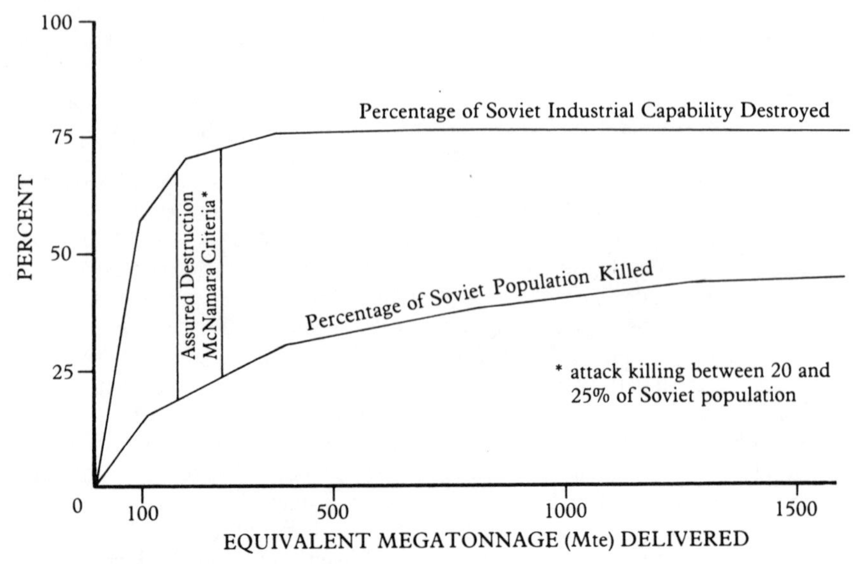

Fig. 10 The Diminishing Returns of Overkill

use of a Hiroshima-sized weapon upon any population. What the Rand model does do, however, is to give the blast deaths per yield for a given area with a given density of population, and by so doing provides the basis of a reasoned forecast that is independent of local conditions such as prevailing weather, terrain and dominant types of construction. By predicting fatalities from blast overpressures in specific areas, the model tends, however, to overstate the number of actual weapons required to do the damage.

US Defense Department data has shown that an attack on the Soviet Union of close to 2000 megaton equivalents would kill roughly half its population and destroy two-thirds of its industrial potential, but the vast majority of this damage would be accomplished with the first 100 to 200 Mte. After this, there are only marginal increases in death and damage for large increases in megatonnage. This is in part what is meant by 'over-kill'. It is true arithmetically that the world's nuclear weapons are enough to kill everyone on earth several times over, but death and destruction do not grow proportionately with the megatonnage used and only a relatively small number of nuclear weapons is enough to cause unacceptable damage on any imaginable scale. The rest exist principally as a redundancy on the assumption that a large number of missiles would be destroyed before being used or for their counterforce capability. The worst official forecasts for what would happen in an all-out nuclear war show up to half the populations of both the USA and the USSR surviving, although they seldom detail the type of world they would inherit.

Comparing deliverable megatonnage is not easy because of the complexities of varied delivery systems. A B-52 bomber, for example, may carry a wide mix of gravity bombs (of various yields) and air-to-ground strategic missiles such as

SCRAM and the ALCM. If the aircraft was equipped with two high-yield (say 10-megaton) gravity bombs in its forward bay and carried its maximum of 20 SCRAM/ALCMs in the aft bay and on external pylons, the total would be 24 megatons for 22 separately targetable weapons. The total amount of equivalent megatons would be 6.3 Mte for the gravity bombs and 6.8 Mte for the 20 SCRAM/ALCMs. Each bomber would thus be capable of carrying 13.1 Mte and the entire force, if armed exactly the same, would theoretically represent well over 4000 Mte.

Adding to its bomber advantage are the American F-111 and F-4 fighter-bombers based in Europe and the US Navy's fleet of carrier-based aircraft in the Pacific and the Mediterranean, all of which are capable of limited attacks on the Soviet Union. The importance of these forces to the strategic nuclear balance is widely disputed but it is clear that they do contribute to a significant degree and that America's bomber-deliverable megatonnage is increased by a useful amount. The dispute centres mainly upon the ability of these forces to penetrate Soviet air defences and their maximum range of operation within the USSR. Of these forces the chief threat is probably that posed by the F-111s based in Europe. Armed with six SCRAMs, an F-111 operating from Britain or West Germany could carry a total of 2 Mte into the western Soviet Union. Because of all the uncertainties involved, these forces are normally left out of strategic comparisons but their importance is significant in an appreciation of the overall picture.

At the beginning of 1981, the balance at sea was in Moscow's favour with up to 1552 Mte for her SLBM fleets. It is in the ICBM category that the Soviet Union clearly dominates in equivalent megatonnage. American intercontinental strategic missiles show a total of 1274 Mte compared to 6013 for the Soviet Union. The Russian figures result from widely varying estimates for the number of warheads carried by the MIRVed SS-18 and their individual yields. The missiles under development will not drastically affect the balance.

The questions of unacceptable damage and deliverable megatonnage for minor nuclear powers are greatly enhanced by the fact that the available Mte is much less. Britain's four Polaris submarines, for example, carry a total of only about 65 Mte but only one sub is guaranteed to be on patrol at any given time. Thus the available megatonnage may be as low as 16.2 which, including non-blast deaths, would be sufficient to kill between 10 and 20 million Russians. The lower figure is just half the losses suffered by the Soviet Union during World War II but Britain's defence planners argue that such a 'small' capability proves an effective independent threat.

'Limited' Overkill

Forecasts for limited nuclear war vary widely according to the scenario adopted. The most important are those involving strikes at military installations but which are also aimed at minimizing civilian deaths so far as it is possible or practical. Cities are thus not targeted and casualties result primarily from military personnel on-site, civilians living on or near the installation and civilians living downwind and exposed to fallout. (Downwind, of course, can mean several hundred kilometres away and include major urban centres.)

In 1974, Secretary of Defense Schlesinger presented the Senate Foreign Relations Committee with data for casualties resulting from a number of different Soviet strikes on military installations within the United States. These were:

Target	**US Deaths**
Land-based ICBMs	1,000,000
Strategic Air Command bases	500,000
SLBM bases and other Naval installations	200,000

The numbers would increase by a factor of two to three if ground-bursts were used instead of air-bursts, which involve less fallout. These figures caused immediate controversy and critics offered alternative forecasts which — including a revision from the Defense Department — brought the totals closer to 20 million rather than the maximum of 2 to 5 million suggested by the above.

In the revised schemes, an attack by two 3-megaton warheads (one air- and one surface-burst) on each American ICBM silo would cause some 18.3 million deaths and destroy something like 80 per cent of the missiles. A similar attack using two 550-kiloton warheads per silo would kill 5.6 million but destroy only 42 per cent of the ICBMs. An all-out strike on America's ICBM, SAC bases and ballistic missile submarine bases, on the other hand, would involve 16.3 million fatalities but succeed in eliminating only 57 per cent of the silos, 60 per cent of SAC's on-ground aircraft and 90 per cent of the missile submarines in port. This attack was assumed to involve two 1-megaton surface-bursts on each ICBM silo and air-bursts over each of the 46 SAC bases and the two ballistic submarine bases. Factors which would alter the results were specific assumptions of Soviet missile performance and certain 'population protection' factors. These refer to civil defence measures which the United States has largely ignored for over two decades. Some European countries, however, have placed considerable emphasis on protecting their populations, and both the Soviet Union and China have reportedly extensive civil defence programmes but their full extent — or importance — is disputed. Recent improvements in Soviet missile performance would not significantly alter the damage estimates except for a far greater proportion of ICBMs destroyed. The numbers of surviving bombers and submarines are likely to be very similar while the amounts of deliverable megatonnage required would be roughly the same. The forecast deaths are largely a result of drifting patterns of fallout which is unaffected by missile performance.

Missile Reliability

No piece of equipment works perfectly all the time. Missile submarines cannot be constantly at sea but require regular maintenance and extensive systems checks. So do the ICBMs resting in American or Soviet silos. There are also unforeseen accidents, such as the 1980 explosion in a Titan silo near Damascus, Arkansas. Analysis of maintenance schedules and the likelihood of accident or system break-down provides an 'unavailability factor' or that proportion of the system which is expected to be inactive at any given time. Of their total forces, as few as 75 per cent of either American or Soviet ICBMs may be available on any given day. Careful scheduling and good system design can maximize availability but it can never make it anything like 100 per cent.

Even if it is available, a missile may not work perfectly when it is fired. Instead it may fail at any one of a number of points from launch to a warhead's arrival at the target. Something can go wrong during the launch, boost, second or third stage separation or ignition, warhead separation, re-entry and detonation. This level of reliability has to be assessed, and is usually symbolized as R, which expresses the estimated chance of the missile functioning perfectly (see note 5). For a number of missiles with the same reliability, R is also that proportion of the total which would be expected to perform successfully. An R of 90 per cent, for example, means that any individual missile has a 9 in 10 chance of a successful performance and 1 chance in 10 of failure. A reliability of 90 per cent is a probability rather than an exact prediction.

In the sense that it is used in strategic thinking to express reliability, 'probability' here is an observed frequency — a proportion of successes over a number of trials interpreted as a prediction of future performance. It is inferred that a missile which has had a 90 per cent success rate in a long-running test series will perform about as well in the future. But ballistic missiles have never been tested under fully operational conditions, which makes these mathematical probabilities more problematic than one might first suspect. The high proportion of successful space shots over recent years leads most people to believe that missile technology is an exact and predictable technology; but the amount of advance preparation, repeatable checking procedures, rescheduling and massive manpower and machine back-up provided for space launchings is not available for an ICBM complex which is constantly on alert and would be called upon to perform almost instantly at a moment's notice. Measuring missile performances in terms of probability is useful, but only so long as the over-simplications are kept in mind and scientific language is not taken as direct evidence of scientific fact.

Explaining missile performance in terms of the mathematics of probability has one bizarre consequence. With a reliability of 50 per cent, a launch of six missiles would be expected to show a 50/50 division of successes and failures. But this is only the *most expected* outcome, and it is perfectly possible to get any one of six other possible results. There might be six successes in the six attempts, for example, or no successes whatsoever. Each of the seven possibilities from total failure to a six out of six success rate has its own unique probability. In principle, it is no different than wagering on getting three heads in six tosses of a fair coin, but the probability of this happening is only 31 per cent rather than the 50 per cent chance one might expect.

The one Polaris submarine Britain guarantees to be on patrol at any one time carries 16 missiles. Should the sub be called upon to fire, those 16 missiles would be launched at targets in western areas of the Soviet Union with an expected proportion of successes equal to the missile's reliability R. Should R be as low as 66 per cent, the predicted result is that 10 or 11 Polaris warheads would reach their targets and cause at most 12 million deaths. But there are 16 other possibilities with unique probabilities running from a virtually impossible chance of 16 failures to an extremely small one of 16 successes.

Because each of these possibilities involves a set amount of equivalent megatonnage delivered on Soviet cities with its own maximum number of deaths, more or less probable fatalities are forecast. Moscow can thus assess an independent British deterrent in terms of its probable threat rather than as one which simply offers a yes or no choice of unacceptable damage. In evaluating policy goals, a high probability for a given level of damage may make it clearly unacceptable; but if the chances are rather low, the risk may begin to seem small and in some sense acceptable. In a Polaris A-3 launch, there is close to a 25 per cent chance that no more than 9 of the 16 MRV warheads would reach their targets and nearly a 45 per cent chance that no more than 10 would do so. At the very worst, this would mean a 25 per cent chance of no more than 9 or 10 million Soviet deaths and a 45 per cent chance of no more than 11 or 12 million.

Polaris is almost certainly more than 66 per cent reliable and any such ambiguities in the British deterrent that still exist will largely disappear when Trident enters service. Instead of 16 warheads (MRVed systems being counted as one), Trident's 14 MIRVed warheads mean that a single submarine will carry 224 individually targetable re-entry vehicles with a total of 63.2 megaton equivalents, giving Britain nuclear superpower status.

During the 1980s China's new CSS-X-4 ICBM is expected to enter medium to full-scale deployment. It will join an obsolete fleet of up to 100 intermediate missiles and around 70 strategic bombers based on the Russian Tu-16. For its first years of service, the CSS-X-4 will have an involved and relatively lengthy launch time and be emplaced in sites with little or no protection against nuclear attack. In the language of the defence community, the entire force has a 'minimum survivability'; little of it would be expected to remain following any serious Soviet or American effort to take it out. As the CSS-X-4 is likely to be fairly unreliable for some time — a generous estimate for reliability is 66 per cent — Moscow could evaluate the possible consequences of a strike in light of the expected surviving missiles and the probable retaliatory casualties derived from intelligence estimates of Chinese missile reliability. An estimated R of 66 per cent and an expectation of no more than 16 surviving ICBMs, for example, would be looked at in exactly the same way as the Polaris example because similar probable fatalities are involved. The real relevance of these assessments of probable damage, however, is that they could be regarded as justifying a strike in line with saving life, because removing the Chinese threat could be rationalized as reducing its capability to harm the Soviet Union and a good chance of escaping with no more than 10 or 11 million deaths. The 'logic' of a damage-limiting pre-emptive strike if war appears likely will, given her strategic weakness, be a great constraint on China's foreign policy for years to come.

Accuracy and Probability

However reliably a missile works, its warhead may still miss its target due to inherent inaccuracies imposed upon the guidance system by the laws of physics. The accuracy of a warhead, as distinct from missile reliability, is measured according to what is known as a circle of equal probability or CEP (see note 6). The CEP is the circular area around the target within which the warhead has a 50 per cent chance of landing, and, like the missile's reliability R, the warhead's CEP value is derived from repeated testing and may be improved through better guidance designs. Accuracy is particularly important in hard targeting because of the more precise requirements produced by the increased blast resistance in the target.

The area around a small, hard target within which the warhead must land in order to destroy the target is known as the lethal radius, or LR. It is calculated from the target's blast resistance in pounds per square inch and the attacking re-entry vehicle's yield in megatons (see note 7). When CEP and LR are identical, the target has a 50 per cent chance of survival or destruction. This is an 'all or nothing' approach — sometimes known as the 'cookie-cutter' — in which it is assumed that the target will be destroyed if, and only if, the warhead detonates within the LR but will survive if the warhead misses.

More sophisticated approaches allow for a target surviving even where the lethal radius is hit or being destroyed by a near-miss. These calculations take into account such things as the predominant terrain and geological features prevailing at the target as well as the duration of the blast wave. The 'cookie-cutter', by contrast, takes only the peak overpressure into account — the peak overpressure being the maximum amount of overpressure generated irrespective of the length of time that the whole wave affects the target. As a rough rule of thumb, the higher the yield the longer the duration of the blast wave and the greater the chances of damage. This gives an advantage to the Soviet Union's larger-yield weapons in point targeting.

A warhead's final kill probability (Pk) can be calculated from the target's LR, the warhead's accuracy in CEP and the attacking missile's reliability R (see note 8). Pk is the single-shot kill probability or the chance that a single warhead will destroy the target. Equally, the chance of a target surviving an attack by one warhead (Ps) is the kill probability subtracted from 1. Where a number of different warheads attack one target, there are simple formulae (note 9) for working out the overall chance of destroying it. This is known as the many-shot or n-shot kill probability (Pk_n). Where single warhead launchers are involved, the exercise is quite straightforward but it becomes slightly more complicated when MIRVed launchers are used (see note 10).

Problems arise when more than one warhead from the same MIRVed launcher are sent to the same target. This reduces the overall n-shot kill probability by marginal amounts which, over a large number of cases, could be quite significant. The solution lies in what is called cross-targeting. Cross-targeting involves maximizing kill probabilities by assigning each warhead of a MIRVed launcher to a separate target whenever possible. For example, to gain the best possible overall kill probabilities for three Minuteman III missiles in an attack on three separate targets, the three warheads on each missile would be assigned differently. Each target would then receive one warhead from each of the three attacking missiles and no target would receive more than one from the same launcher. This is a simple example of cross-targeting, but handling a large number of targets and missiles would be far more complex.

Although increased yield boosts kill probability, accuracy is more important. Improvements to Minuteman III's accuracy and yield have effectively doubled the missile's chances against harder targets, and recent improvements to Soviet missiles give excellent kill probabilities once the higher yields of their warheads are taken into account. Yet accuracy is still more critical than yield. For example, a 340-kiloton warhead with a CEP of 182 metres (600 feet) has approximately a 5 per cent better chance of killing a 1000 psi silo than a warhead with a yield of 2 megatons but a CEP of 365 metres (1200 feet), despite the fact that the latter carries nearly six times the yield.

Once a certain level of accuracy has been reached, however, higher yields can add significantly to kill probability. (Soviet guidance technology has now reached this point and, although their warheads are less accurate than their American counterparts, Soviet missiles have comparable kill probabilities.)

Kill probabilities can also be calculated from a warhead's lethality, or K, which expresses a relationship between yield and accuracy. The higher the value for K, the more lethal the warhead. The method of calculating K and deriving kill probabilities from it is given in note 11. Using the lethality formula, it is possible to draw up comparative numerical totals for the respective counterforce capabilities of the superpowers as illustrated opposite. A single improved Minuteman III warhead, for example, has a lethality of 48.7 and, as each missile carries three MIRVed warheads, each missile has a total lethality of 146.1K. As there are 550 Minuteman IIIs deployed, the lethality of the total force would be 80,355K.

Yet comparative lethality tables may be somewhat misleading. A single air-breathing Tomahawk cruise missile will have a lethality of 2783.4K and 29 Tomahawks will have the same total lethality as the 550 Minuteman ICBMs currently deployed. But 29 Tomahawks, carrying only a single warhead each, could attack only 29 targets; the 550 Minuteman III, by contrast, can attack 1650 targets. There can be other sources of misunderstanding too. One Tomahawk will have about 14 times the lethality of one MX warhead but each will be able to attack the hardest of targets with kill probabilities virtually indistinguishable from certainty.

During the 1980s both superpowers will deploy missile systems which possess combinations of accuracy and yield which will give kill probabilities approaching 100 per cent. In theory, then, a one-warhead-one-target relation-

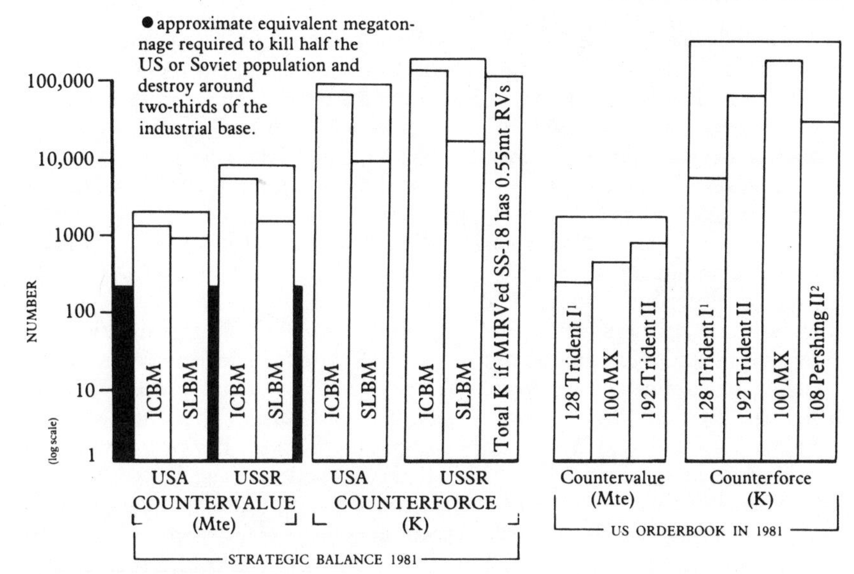

Fig. 11 Strategic Balance — Force

Excludes aircraft-delivered megatonnage which would favour the US. Not enough is known about Soviet plans to quantify the order book in any detail.

1. Before the Trident II go-ahead, the order book called for 320 Trident I and this may still be the final figure.
2. Based in West Germany. Cruise missiles are not compared because the numbers are uncertain.

ship would be the counterforce 'ideal' as there would be no need to increase kill probability by sending more than one warhead to any target. This overlooks the reliability problem, however, since nothing would be accomplished if the warhead failed to arrive or to detonate. A Tomahawk or MX warhead with a 99.9 per cent chance of killing its target and with a system reliability of 85 per cent will have only an 84.9 per cent final kill probability for a single attempt. While this is impressive, it does not amount to certainty, and it would be necessary to send more than one warhead to push the final kill probability to something approaching 100 per cent. As new systems become more reliable through repeated testings and technological improvements, reliability may creep into the middle 90th percentile and perhaps even begin to approach certainty itself. When that happens, the one-warhead-one-target guideline would become realistic, but that is at least 15 years away.

The idea of increasing kill probability by sending more than one warhead to a target is intuitively reasonable but certain characteristics of nuclear explosions spoil its theoretical purity. Nuclear explosions create blast waves, radiation, thermal and electromagnetic effects. They also create severe wind storms and high concentrations of débris and dust. The ionization of atoms in the atmosphere creates rapidly oscillating electrical and magnetic fields, known as the electromagnetic pulse or EMP, which are highly destructive to the sensitive electrical and electronic equipment of a missile's on-board guidance and control system as well as those of its command and launch network.

Any of these nuclear effects may damage other incoming warheads. EMP or radiation may deactivate a warhead, for example, or blast or severe wind deflect it. A first shot which failed to kill its target may thus interfere with a second or third warhead. This is known as the interference or 'fratricide' effect and makes the simple mathematics of increasing kill probability by using more than one warhead dubious in practice. In theory, it is possible to minimize the uncertainty introduced by fratricide by carefully timing the arrival of incoming warheads and taking advantage of the 'pin-down' effect. But in practice this would be difficult. The second or third warhead would have to arrive at the silo after the probability of interference had dropped to tolerable proportions but before the attacked missile could be launched and reach a safe altitude.

Following an attack, it would be important to assess the results through satellite reconnaissance to determine which silos had survived and which were destroyed. But this would also be hindered by the dust and débris created, so that any second strike on the survivors would almost certainly either have to be delayed or be largely guessed at. Although it is possible to reduce the need for second strikes, fratricide cannot be eliminated from the calculations until both missile reliability and warhead accuracy are sufficiently high to give a single re-entry vehicle a kill probability approaching certainty.

MX should have the necessary accuracy, but when its reliability is taken into account it will still be essential to use a second warhead to take its kill probability close to certainty. The same is true for the new cruise missiles, although their planned accuracy is far more precise. Cruise missiles also suffer from slowness and from the fact that their evasive abilities could mean that a launch of them could be taken as an attack on cities instead of a counterforce strike.

A degree of inaccuracy can be offset in assessing silo-killing ability by the use of 'earth-penetrator' warheads. Conventional earth-penetrating techniques have been used for years in bombs aimed at such targets as airfield runways, and usually the technique involves slowing the bomb's descent by parachute and, in the final seconds, driving it into the ground with an explosive charge. It is quite possible, but not certain, that Pershing II can be equipped with a nuclear earth-penetrator, perhaps designed for the incoming re-entry vehicle to fire a sub-projectile into the ground just prior to impact to increase the depth of penetration. Using sub-surface explosions, a nuclear weapon's cratering effects may be as much as five or six times that expected from an equivalent yield detonated above ground. In effect, the weapon would have a chance of uprooting a buried silo even if it should survive the blast as heavy materials in the warhead would probably contribute to an increase in the seismographic shock created by the explosion. The earth tremors created by nuclear weapons can shift the ground up to a metre or so anyway, and increasing the earthquake effect would increase the chances of damaging the silo.

1. *(previous page) Shape of a ladder permanently etched into the wall of a tank by the atomic bombing of Hiroshima.*

2. *Distinctive mushroom-shaped cloud formed by the explosion of an atom bomb.*

3. *The immense cloud formed by the detonation of a hydrogen bomb seen from several miles' distance.*

4. *Prototype B-1 bomber.*

5. *B-52 strategic bomber. This one is armed with air-launched cruise missiles.*

6. *The controversial Soviet Tupolev TU-26 Backfire medium-range bomber.*

7. *Impression of the 'invisible' Stealth bomber proposed for deployment in the 1990s.*

8. *America's ageing Titan II intercontinental ballistic missile.*

9. *USS Michigan, an Ohio class submarine equipped with twenty-four Trident missiles.*

10. *The Polaris missile (left) with its MIRVed successor, the Poseidon.*

11. *The French MSBS M-20 submarine-launched missile.*

The Minuteman III
ercontinental ballistic
ssile.

America's newest
marine-launched missile,
Trident, in a test launch at
be Canaveral. This missile
lso being bought by
tain.

14. *Impression of a Soviet SS-20 missile.*

15. *Full-scale mock-up of the super-accurate MX intercontinental missile.*

16. *The Pershing II medium-range missile, scheduled for Western Europe in the mid-80s.*

17. *The 'Sprint' anti-ballistic missile.*

18. *The air-launched version of the ground-hugging 'intelligent' American cruise missile.*

19. *Photograph and design of the first cruise missile, the German World War II V1.*

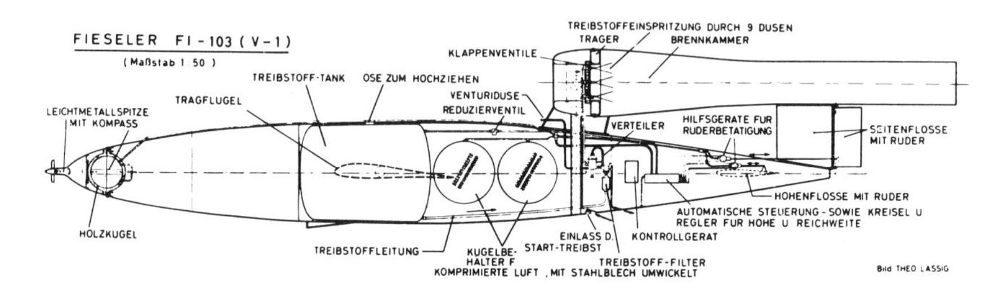

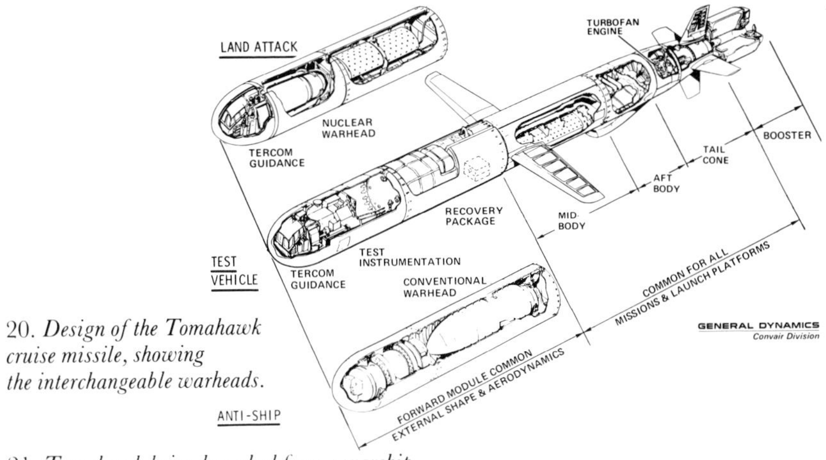

20. *Design of the Tomahawk cruise missile, showing the interchangeable warheads.*

21. *Tomahawk being launched from a warship.*

22. *Diagram of an umbrella-gun as thought to be used in the assassination of Georgi Markov.*

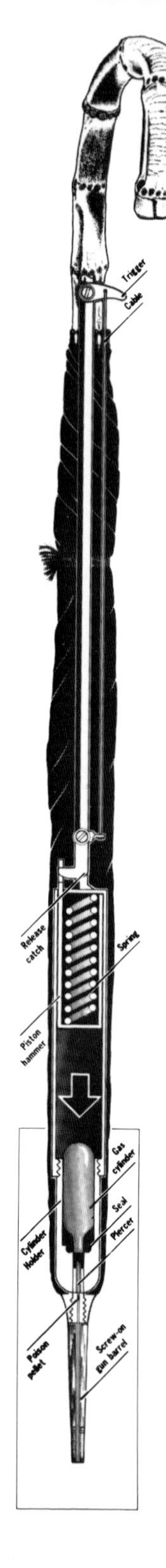

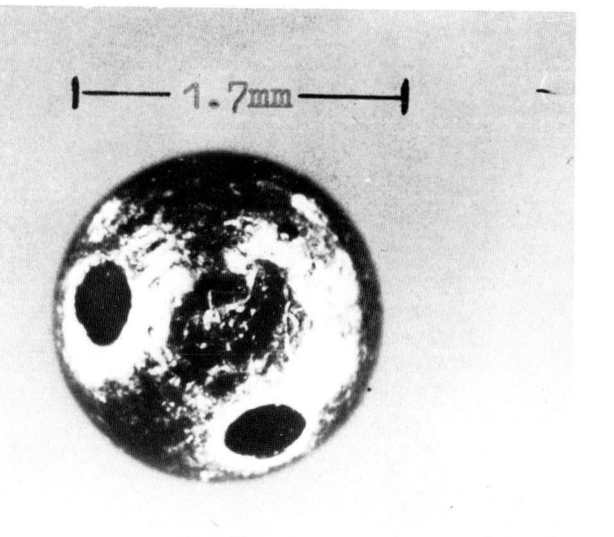

23. *The minute platinum-iridium pellet found in Markov's body. The two holes drilled in at right angles carried the lethal toxin.*

24. *Pretoria's route to the Bomb? A South-African uranium-enrichment plant.*

25. *Impression of a Soviet laser weapon, used to destroy incoming missiles or planes. The second vehicle is a power plant.*

26. *Gas shells being prepared for firing in World War I.*

27. *The effects of anthrax on a human arm.*

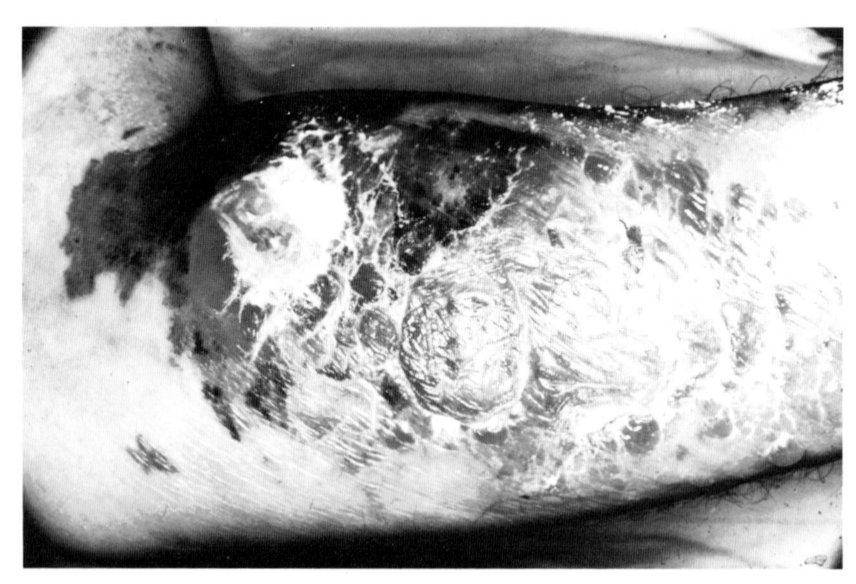

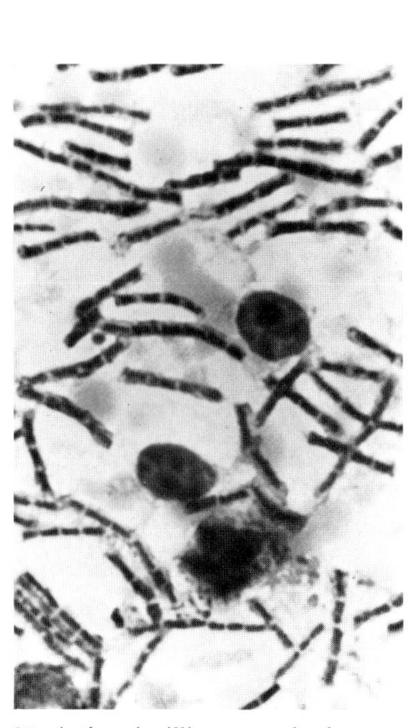

28. *Anthrax bacilli as seen under the microscope.*

29. *Five of 5000 sheep accidentally killed in 1968 by a nerve gas test in Utah.*

30. *Military policeman guarding a train carrying nerve gas rockets in Bremen, Georgia.*

31. *A NATO infantryman wearing special clothing designed to protect against the effects of nuclear, biological or chemical attack.*

It would be possible to use Pershing II in conjunction with ICBM warheads to increase kill probability. An earth-penetrator would confine the effects of Pershing II's low-yield explosion to a degree that would reduce to a minimum its possible interference with a second incoming warhead. The second warhead, fired by an ICBM or SLBM, would be timed to arrive at the same time or just after Pershing II and could be set for an air-burst to reduce the chances of fratricide even further. This technique could overcome any problems with Pershing II's reliability and give a high kill probability with greatly reduced chances of the two warheads interfering with each other. Using a combination of one Pershing II warhead and one MX warhead, for example, could give a 97.7 per cent chance of killing a 2000 psi target after the reliabilities of the two missiles were taken into account.

The New Missile Gap?

When John F. Kennedy made his bid for the presidency in 1960, he exploited a widespread fear that the Eisenhower administration had allowed the Soviet Union to take a dangerous lead in developing intercontinental ballistic missiles. There was a barrage of warnings that America faced years of Soviet intimidation and international adventurism until the United States closed this ICBM 'gap'. Soon after Kennedy's election, it became clear that the missile gap did not exist and that most of the alarm had come from misreading Soviet military capabilities and taking Russian statements of rapid advances in ICBM technology, production and deployment at face value. Despite official denials, the missile gap was widely believed and probably contributed greatly to Kennedy's narrow victory.

If Pentagon statements are to be accepted, the United States is now in the middle of another missile gap. This time it does not involve numbers of missiles but rather a period in which America's ICBMs are vulnerable to a Soviet strike which, according to official statements, would be capable of destroying 90 to 95 per cent of them by the mid-1980s. This so-called 'window of vulnerability' (the official phrase) will supposedly exist until the mobile MX enters full service or some other means of protecting Minuteman is found.

Using the formulae for missile reliability and warhead kill probability, it is possible to gain a rough idea of each side's counter-silo capability, despite the complexities of the variables involved. Even the generally reliable estimates of yield and the numbers of MIRVed warheads carried on certain missiles can vary widely, especially for the Soviet Union. Nevertheless, the exercise is worthwhile because it puts the ICBM vulnerability argument into a concrete format and illustrates some of the problems involved.

Should Washington wish to take out all 1398 Soviet ICBM silos with its own ICBMs, it would be forced to draw solely upon Minuteman and Titan until MX comes on-stream. In line with having a simplified look at respective counterforce capabilities, the contribution of manned bombers, cruise missiles and submarine launched missiles will be left out. American SLBMs will have little silo-killing capability until Trident II enters service in around 1990. Even

Fig. 12 Strategic Balance — Numbers

	USA	*USSR*	*UK*	*France*	*China*	*Europe*[1]	*Warsaw Pact*[1]
ICBM	1052	1398	0	0	4	0	0
I/MRBM	0[5]	610	0	18	75		500 +
SLBM	576	989[2]	64	80	0	184[3]	30 +
(missile subs)	36	84[2]	4	5	0		
Bombers							
Long-range	316	150	0	0	0		
Medium-range	60	500	55	46	100 +		
Dual-capable aircraft assigned a nuclear role[4]						1160	3095
Nuclear artillery and short-range missiles (nuclear warheads)						1400	1600

1. These figures are intended to give only a rough idea of the balance of theatre nuclear forces in Europe. Arms negotiators disagree both over numbers and on definitions of 'theatre', thus the West maintains that Britain's and France's nuclear missiles are strategic rather than theatre weapons.
2. US Defence Department estimate: 950 SLBM in 62 subs.
3. Including about 40 US Poseidon SLBM assigned to NATO. These have been counted as strategic in previous arms negotiations.
4. An ISS estimate of the approximate numbers of aircraft which would be given a nuclear rather than conventional role in a European war.
5. Officially the US intends to station 108 Pershing II and 464 Tomahawk cruise missiles in Europe but the thousands of Tomahawk and ALCM on order for US forces will confuse European arms negotiations because of the simplicity of using them in theatre roles.

if it were perfectly reliable, a single Poseidon warhead would have only an 11 per cent chance of destroying a 500 psi silo and it would take seven of them to boost the chances of a kill to around 55 per cent. Trident has a better chance with 33 per cent for a single warhead but this is still a long way from a true silo-killing potential. Although a real attack would probably involve elements from the entire strategic armoury, the intricacies of counterforce targeting and the complexities of the new 'missile gap' can best be clearly explained by basing it upon the improved Minuteman III alone.

America's 550 Minuteman IIIs carry a total of 1650 MIRVed warheads which, assuming perfect cross-targeting, could be distributed evenly on 1398 Soviet missile silos. Since there are 252 more warheads than targets, 252 Soviet silos would receive two warheads each and the remaining 1146 one each. This is the optimal assignment only so long as each target has the same value to the attacker which, in practice, is unlikely. The newer and more accurate Soviet MIRVed SS-18s, for example, would certainly have a higher priority in American plans than the older, inaccurate SS-11s.

Calculating Minuteman III's final kill probability from its CEP and yield (see table on pages 68-9) and an assumed 90 per cent reliability, the attack would be expected to destroy between 83.3 per cent (1165) and 91.5 per cent (1279) of Russia's ICBMs. The two figures represent different assumptions of the hardness of Soviet missile silos; in this case, 1000 and 100 psi respectively. Reports of Soviet silo upgrading suggest the lower percentage would be more likely but if rumours are true that super-hard silos of several thousand psi are being built, it would reduce Minuteman's capabilities well below these figures.

However, Minuteman's performance would actually be better than the percentages given because of the already existing facility for reprogramming. This is a system which enables on-board guidance instructions to be quickly altered to assign the missile to a new target. The rapid switching of targets — from cities to silos, for example — can be undertaken in a matter of minutes as the politics of the moment dictate, thus allowing a shift from a counter-city to a counterforce posture, or vice-versa. Until the late 1970s retargeting was a laborious procedure which required the physical insertion of a new set of encoded tapes into the missile's guidance package, which took between 15 and 24 hours. To increase Minuteman III's flexibility, the United States developed what is known as the command data buffer system which enables retargeting via deeply buried data lines linked to the five launch control centres (LCCs) of a 50-missile Minuteman squadron. There is one LCC for every ten missiles but every missile is linked to each LCC. This allows any LCC to replace any other lost through malfunction or enemy action. Each LCC is located several kilometres from any other and from any missile silo; each is 7.5 metres (25 feet) long by 1.5 metres (5 feet) wide and is buried some 21 metres (70 feet) below ground. The entire system is interactive. The missile signals that it has received new instructions and informs the LCC when the retargeting is complete. Instead of up to a day, retargeting can now be completed in less than 30 minutes. Retargeting may also be carried out from an airborne launch control system (ALCS) which is designed to duplicate ground-based LCCs and replace them if necessary. Usually a modified Boeing-707, the ALCS takes off upon an attack warning so as to escape destruction upon the ground and may remain aloft for about a day.

Reprogramming effectively increases missile reliability and, therefore, kill probability. It can do this because, when a missile fails, it is possible to reassign another to the same target and launch it to fill the gap. For example, out of a launch of 100 missiles with a reliability of 90 per cent, 10 would be expected to fail. An attack using reprogramming would launch the first 100 missiles at their targets and hold another 10 in reserve to replace failures as they occurred. Obviously it must be possible to know which missiles actually failed and which did not. When reprogramming is used, the missile's reliability R may be looked upon as its reliability during the stages of flight after which a failure cannot be identified. This is known as a missile's 'non-reprogrammable reliability'. Its 'reprogrammable reliability' is its reliability from launch for the period that failures can still be identified. The details are found in note 10 but, in brief, the number of missiles required for the reprogrammable reserve force is found by calculating the number in the initial launch which are expected to fail during the period in which failures can be identified. Obviously, if all failures could be identified and reprogramming was sufficiently rapid, missile reliability would no longer be a critical variable in calculating kill probability.

LAND-BASED

	Missile	Number	Range (nautical miles)	RV	Yield per Re-entry Vehicle (megatons)	COUNTERVALUE (Mte) per Re-entry Vehicle	COUNTERVALUE (Mte) per missile	COUNTERVALUE total	CEP[7] (nautical miles)	COUNTERFORCE Pk(%)[3]	COUNTERFORCE K missile	COUNTERFORCE K total
USA ICBM	Titan II	52	8100	1	9	3	3	156	0.5	51.7	'7.3	900
USA ICBM	Minuteman II	450	6100	1	1+	1	1	450	0.3	37.3	11.1	4995
USA ICBM	Minuteman III	250	7000	3	0.17	0.306	0.92	230	0.125	43.7	58.8	14,700
USA ICBM		300		3	0.34	0.487	1.46	438	0.1	87.1	146.1	43 830
USA ICBM	MX	0	7000+	10	0.34	–	–	–	0.05	99.9	–	–
USA	Tomahawk (cruise)	0	1500	1	0.2	–	–	–	≤0.01	99.9	–	–
USA	Pershing II	0	1000+	1	∼0.005	–	–	–	≤0.01	99.9	–	–
USSR ICBM				1[4]	1.5	1.22	1.22	708	0.5	19.8	4.89	2836
USSR ICBM	SS-11	580	5700									
USSR ICBM				3[1]	0.15							
USSR ICBM	SS-13	60	5400	1	1	1	1	60	0.4	23.1	6.25	375
USSR ICBM				4[4]	0.9	0.932	3.73	559	0.3	35.3	41.4	6210
USSR ICBM	SS-17	150	5400									
USSR ICBM				1	5							
USSR ICBM			5700	1	24				0.22≥	99.9		
USSR ICBM	SS-18[2]	308	5000	8[2, 4]	2	1.41	11.3	3480[2]	0.17≤	90.1	391.4	120,551
USSR ICBM			5700	1	1				0.09≤	99.9		
USSR ICBM			5000+	1	10				0.1≤	99.9		
USSR ICBM			6000	6[4]	0.55					62.5		
USSR ICBM	SS-19	300	5500	1	5	0.67	4.02	1206	0.17	98.5	139.3	41,790
USSR	SS-4	340	1000	1	1	1	1	340	0.6	11	2.8	952
USSR	SS-5	40	2200	1	1	1	1	40	0.5	15.5	4	160
USSR			2700	1	1.5				0.22≥	68		
USSR	SS-20[6]	∼250	3000	3[4]	0.15	0.282	0.846	212	0.2≥	25.7	7.06	1765
USSR			4000	1	0.05				0.17≤	17.9		
USSR	SS-22	?	539	?	?	–	–	–	?	?		

	Missile	Number	Range (nautical miles)	RV	Yield per Re-entry Vehicle (megatons)	COUNTERVALUE per Re-entry vehicle	(Mte) per missile	total	CEP (nautical miles)	COUNTERFORCE Pk(%)[3]	K missile	total
USA	Polaris	80	2500	3[1]	0.2	0.71	0.71	57	0.5	11.2	2.84	227
	Poseidon	432	2500	10	0.05	0.135	1.35	583	0.3	6.1	15.1	6523
	Trident I	64	4000	8	0.1	0.215	1.72	111	0.2	20.3	43.2	2765
	Trident II	0	6500	14	0.15	–	–	–	∼0.1	69.5	–	–
USSR	SS-N5	57	1500	1	1.5	1.22	1.22	69	1.5	2.4	0.58	33
		165	1300	1	1.5	1.22	1.22	201	1≤	5.3	1.31	216
	SS-N6	288	1600	3[1]	0.3	0.932	0.932	268	1≤	3.8	0.932	268
	SS-N8	291	4300	1	1.5	1.22	1.22	355	0.5	19.8	5.24	1525
	SS-NX17	12	2700	1	1.5	1.22	1.22	15	0.25	41.9	20.9	251
	SS-N18	176	4500	3	1.5	1.22	3.66	644	0.22	68	81.3	14,309
	SS-NX20[5]	0	4200	>3	?	–	–	–	?	–	–	–

Fig. 13 Strategic Missiles — Numbers and Performance

1. MRV system. The total MRV yield is counted as 1 warhead.
2. Another frequent estimate for the MIRVed SS-18 is 10 (or more) warheads of 550 kilotons each. For the SS-18 force, this means a total Mte of 2066 and a K of 71,456.
3. Chance for a single warhead to kill a 1200 psi target with perfect missile reliability.
4. Because there is a wide variety of possible models, the number in italics represents the one thought to be most prevalent. It is also the model chosen for calculating equivalent megatonnage and K.
5. The Soviet SS-NX20 was developed for the new Typhoon-class missile submarine but is believed to have failed its first sea trials in 1981. According to *Soviet Military Power* it carries 12 MIRVed warheads.
6. According to *Soviet Military Power* 175 of the 250 or so Soviet SS-20s are targeted upon Western Europe. In addition, the book identifies 65 SS-20 sites under construction as of mid-1981 and a possible further 100 to 150 under active consideration.
7. CEP estimates are derived from a number of sources and, since small differences can make improvements to kill probabilities and K figures, the estimate represents the most prevalent quote. Should the US elect to install MARV systems on MX and/or Trident II, their CEPs would drop to around those quoted for Pershing II and Tomahawk/ALCM.

If Minuteman III's reprogrammable reliability were as high as 96 per cent — most missile failures actually occur during the early, identifiable, periods of flight — the outcome of its attack would be the destruction of between 85.8 per cent and 94.6 per cent of Soviet ICBMs. Reprogramming would thus add between 35 and 44 more Soviet ICBMs to the numbers destroyed and bring the total of kills to between 1200 and 1323 without any change in the accuracy or overall reliability of Minuteman. Minuteman's reliability during the non-reprogrammable phase of its flight is assumed to be 94 per cent. In the attack with reprogramming, an initial launch of 528 Minuteman III missiles would distribute their total of 1584 warheads two-to-one on 186 Soviet silos and one-to-one on the remaining 1212 targets. Twenty-two missiles (66 warheads) would be held in reserve for retargeting and launched to replace identified failures. The added kills come from the fact that Minuteman's overall reliability has been effectively raised from 90 per cent to 94 per cent by the addition of the reprogramming facility. In practice, the reprogrammable reserve would be larger than assumed here since it would have to account for failures in the reserve itself and for the possibility of more failures in the initial launch than were predicted.

Assessing the effectiveness of a Soviet attack on America's ICBMs is more difficult to calculate as estimates for such things as accuracy, yield and the numbers of MIRVed warheads are far less certain. This is especially true of the SS-18 which, along with the SS-19, is seen as posing the greatest threat. The graph on page 71 shows a range of possibilities for a Russian attack and some of the varied consequences. It reflects a Soviet concentration upon destroying the 550 Minuteman III silos rather than an across the board attack that treats all American ICBMs as equally important. At the very least, the Soviets would destroy something like 84 to 90 per cent of America's 1054 ICBMs without reprogramming and 86 to 92 per cent with reprogramming. At most, approximately 89 to 95 per cent of American silos would be destroyed. The lower reliabilities and accuracies shown for Soviet ICBMs reflect the position in 1981 but there is evidence that accuracies may be currently somewhat greater than those indicated. If so, the proportions of Soviet kills would increase slightly.

Approximate though these models are, they do bear out the Pentagon's claim that its ICBM force is vulnerable to a pre-emptive strike. Improvements in Soviet missile accuracy and reliability, the Pentagon says, will give the Russians the capability for destroying up to 95 per cent of America's ICBMs by the middle of the 1980s. But there is no 'gap'. What exists is a rough parity in counterforce capabilities with the Soviets possibly enjoying a slight edge. It is the fact that the Soviets have acquired a counterforce parity with the United States to go along with the parity in launchers which they attained years ago that is the cause of concern. There is no clear strategic advantage for either side but rather an equivalence in capability as well as numbers. The fear in Washington is that Moscow might try and exploit its capabilities and attempt to take out America's ICBMs, leaving the occupant of the Oval Office with a

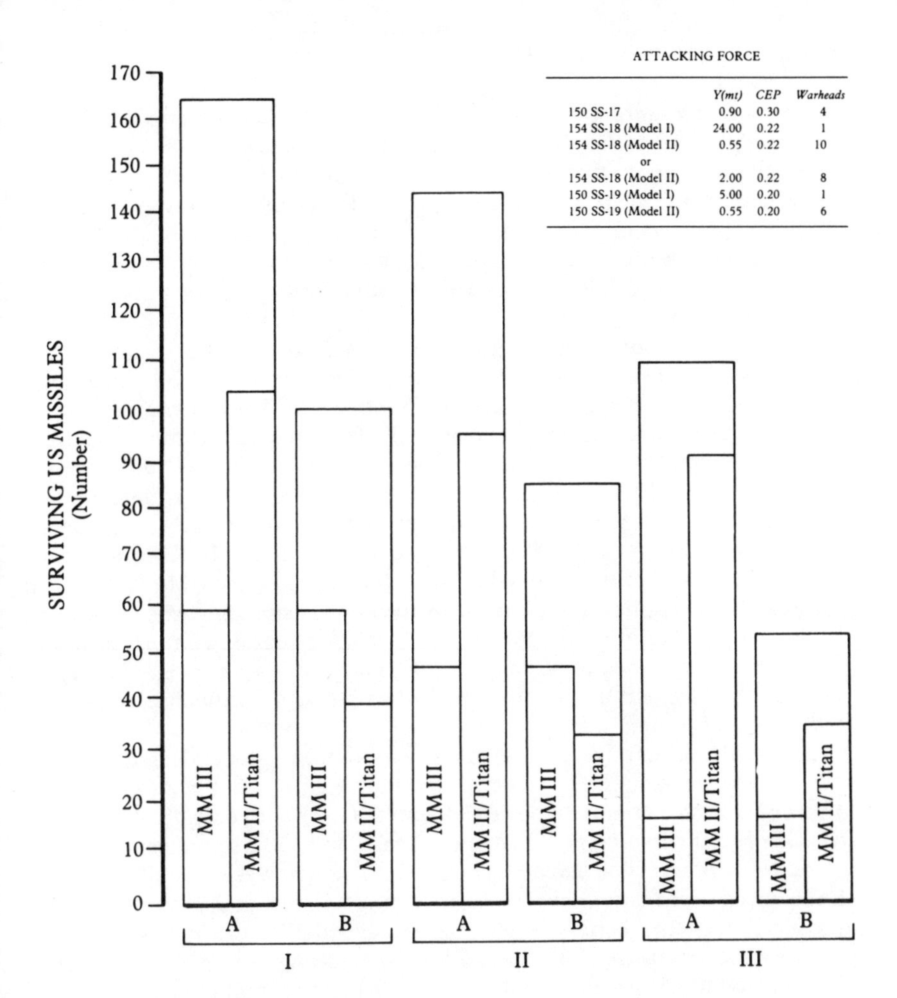

Fig. 14 Soviet Counterforce Capability, 1981

Case

I. a) No Soviet reprogramming
   b) 90% missile reliability
   c) Ten 550kt warheads per Soviet SS-18 Model II

II. a) 85% Soviet reprogramming reliability
   b) 94% Soviet non-reprogramming reliability
   c) Ten 550kt warheads per Soviet SS-18 Model II

III. a) As in II save: eight 2-megaton warheads per Soviet SS-18 Model II

Situation

A. All US silos of 1000 psi
B. 550 US silos of 1000 psi and 504 of 500 psi

choice of doing nothing or retaliating with a city-killing strike using SLBMs and aircraft. This is the justification for the vast expense involved in the MX project — the need for a mobile ICBM combining both an assured survivability and an accuracy precise enough to destroy any Soviet target that American planners might wish to attack. Ultimately, it is also the justification for most of the strategic weapons projects described; the perceived need of the United States to re-acquire a clear strategic superiority in the face of increasing Soviet capabilities.

Assessing the counterforce implications of MX and Trident II is difficult. First of all, the final number of MX is uncertain. Secondly the blast resistance of Soviet silos and the (likely) numbers of mobile ICBM after 1990 is unknown at the moment. However, making reasoned guesses one can calculate the likely results of a hypothetical attack in the 1990s on 1398 Soviet ICBMs emplaced in silos of 6000 psi (400), 2000 psi (800) and 500 psi (198). The attacking force is assumed to be 126 MX (10 warheads per missile), 54 Pershing II (launched from Western Europe), 550 Minuteman III and 48 Trident II. This represents about 20 per cent of the planned Trident force, all the available Minuteman IIIs, half the Pershing deployment scheduled for Europe and just over half of a likely 200-plus MX force ordered before 1990. Non-reprogramming reliability is 95 per cent and the overall reliability R is taken as 85.5 per cent. No reprogramming of Trident II is assumed, reflecting presumed difficulties in keeping tabs on SLBMs after launch.

The results show only 42 Soviet missiles surviving (3 per cent), assuming that the 6000 psi silos house the most 'valuable' missiles, and the attack concentrates on the most important targets. A similar attack with only 50 MX but 120 Trident II, 70 Pershing II, 225 Minuteman II and 550 Minuteman III shows close to 100 Soviet silos remaining. This is still an impressive result but would be achieved at the cost of expending a large proportion of America's strategic missiles. Even the better scenario leaves 42 missiles capable of inflicting catastrophic retaliation on the USA. Convinced that war was imminent, Washington might see its capability as a prudent way to reduce greatly the level of damage it would sustain. But this would be a commitment to disaster, albeit a 'reduced' one, and would not be a preferable option. As an alternative, the selective removal of some Soviet ICBMs is within even present-day capacity, but since Soviet ICBM sites tend to be located closer to major population centres than their American counterparts, even minimal attacks would involve tremendous casualties, so such a war would not remain limited for long.

Counterforce modelling is not an exact science. Missile reliabilities may be lower than allowed for in the graph. No allowance was made for malfunctions within the reprogramming or communications chain. Perfect cross-targeting was assumed but would be highly unlikely in practice. It is equally unlikely that the reprogramming reserve would be able to cover adequately every target left intact because of failure in the initial launch, and the largest intangible, fratricide, was omitted for the sake of simplicity. Moreover, as the USSR becomes faced with its own 'window of vulnerability' later in the decade, it will certainly respond by deploying mobile ICBMs or (possibly) adopting a 'launch-on-warning' policy. In this case, at least some ICBMs are fired in retaliation when the warning system indicates an attack — rather than, as is the stated official policy of both sides at the moment, of waiting until they land. The added dangers for accidental war need no elaboration.

5 Politics and Proliferation

MAD and the Diplomacy of Terror

In the nineteenth century Clausewitz coined his famous maxim: 'War is the continuation of politics by other means.' War, in the sense of military strategy, tactics, logistics and weaponry, has — as Clausewitz also remarked — its own language but not its own logic. Horrific as war is, it is still something fought or avoided for political reasons and as the consequence of political decisions. Specifically tactical or strategic military objectives are generally imposed or restricted by decisions reflecting wider political goals and, while either may change during the course of a war, the fighting is still not something engaged in for its own sake but as a means to a political end — however perverse it might be.

Nuclear weapons have not altered this relationship. What they have done is to change the rules from war as a continuation of politics to the threat of war as a form of political manipulation. At this level, there is no such thing as nuclear 'strategy' in an exclusively military sense. Instead there is nuclear 'diplomacy', and the usefulness of the weaponry consists in its ability to influence a potential enemy rather than its battlefield effectiveness. The overt or implicit threat of military force has always been an adjunct to diplomacy but it is only since the advent of nuclear weapons that discouraging war has been considered the overriding objective. As long ago as 1946, University of California professor and Rand Corporation consultant Bernard Brodie wrote that the primary purpose of the military system was now to avert wars rather than win them. He also noted of the United States and the Soviet Union that an 'equality of deterring power would prove the best guarantee of peace and tend more than anything else to approximate the views and interests of the two countries'.

As Soviet strategic forces caught up with American capabilities in the late 1960s and early 1970s Brodie's view became embodied in the notion of Mutual Assured Destruction (MAD). As a principle, MAD is simplicity itself: if both sides are able to destroy each other even after they have been attacked first, neither is likely to take undue risks, if only out of self-interest. However, there are two essential considerations which complicate the matter. These are capability and credibility.

Capability means having sufficient strategic forces to cause unacceptable damage in retaliation. 'Unacceptable damage' is not itself a fixed quantity but rather an estimate by one side of what the other side is likely to regard as beyond the bounds of tolerance. The analyst Herman Kahn, for example, suggested in 1960 that Americans would 'accept' between 10 and 30 million American deaths as the cost of defending western Europe, but no more.

Credibility refers to the willingness to use the military capability. For a threat to be effective, it must be believed. If Moscow does not believe the American nuclear guarantee to NATO, the threat itself may be ineffective in deterring a Soviet invasion. It is not a case of doubting America's word but rather of believing that the United States would actually act as it promised when the time came. As the Mad Hatter pointed out to Alice, one can say what one means without meaning what one says. Most of the public debates on various weapon systems and strategic policies come down to arguments on credibility. Few intelligent people would pretend that the destructive capabilities of either side are not sufficient to cause appalling levels of damage under any circumstances, but what is disputed is whether certain capabilities are necessary in order to make the threats believable and therefore effective.

As a very loose analogy, imagine two business rivals. If one of them possesses documentary evidence which would totally ruin the other's personal and professional life, he can blackmail him for his own ends. But if both possess such evidence, both are comparatively safe since the threat is mutual and self-cancelling. Both parties are constrained from ruining each other solely on grounds of self-interest. This is the usual picture of MAD: rational restraint out of a shared interest in survival. But MAD is equally an invitation to the two parties to exploit lesser opportunities where it seems apparent that the other might prefer to give way rather than risk a mutual disaster. Quite simply, if it is not credible that the other would risk crossing the brink in order to prevent a marginal defeat, it may be advantageous to be headstrong and adventurous.

The justification of flexible deterrence is to dissuade limited provocations by possessing a capability for limited nuclear responses. This sort of reasoning has been described as the 'logic of the madhouse' and often the assumptions and conclusions of the theories of nuclear deterrence appear more macabre than the weapons they represent. But it is an error to suppose that the doctrine has changed to any substantial degree. What has changed is that new capabilities have opened up new options and the doctrine has been extended to include both. In effect, MAD now encompasses the vocabulary of counterforce and limited nuclear war. Its focus has changed from the threat of an ultimate catastrophe to the threat of a series of mini-disasters which — unless terminated through political negotiation — would lead to mutual catastrophe. This idea has been around for years and Thomas Schelling once called it 'inner-war deterrence'.

If the switch in MAD's emphasis seems illogical and dangerous, it is because the illogicality lies in the original doctrine. The idea that MAD as such is necessarily more stable or more conducive to peace than its more aggressive-sounding extensions is the result of reading one part of the doctrine and ignoring its corollaries. Without flexible or graduated response MAD cannot deal effectively with more than two parties. Take the two business rivals with incriminating evidence on each other. They have a MAD-like stability in their relationship, but if a third competitor joins in, any one man will be able to incriminate either or both of the others. The inherent stability of the one-to-one relationship is reduced because an escalating quarrel between any two automatically endangers the third even if he is uninvolved in the original argument.

Should two cross the brink the third has no guarantee of immunity because neither of the principal antagonists is any longer constrained on the grounds of rational self-interest. When China acquires some capability to target the United States as well as the Soviet Union with ICBMs the MAD relationship will be made more complex than at present and each of the three powers will be intimately involved in the security of the other two.

The central paradox at the basis of nuclear diplomacy is the presumption that to avoid a war it is necessary to appear both able and willing to fight one if unduly provoked. MAD requires that either both parties behave reasonably or that neither appears totally rational. Otherwise, one may exploit the other's fear of disaster by appearing irrationally committed and no longer in total command of events. Borrowing a phrase from the common law, Schelling once called it 'giving an opponent the last clear chance' — creating a situation in which he believes himself to be the only one capable of retaining a reasonable control over events and saving himself.

Because rationality and reasonableness may be a political liability many academic writings on nuclear diplomacy have referred to the theoretical advantages of acquiring a genuine ability to pre-empt choice. If two rival leaders face a situation in which mutual disaster is a possible outcome, there may be an advantage in one leader demonstrating that he can no longer prevent it. The other then has to choose whether to allow the disaster to occur or to give way and prevent it happening. In effect, one makes an irrevocable commitment and leaves the other with Schelling's 'last clear chance'. Strategists have drawn the analogy of a game of 'chicken' in which one of the rivals gets into his car dead drunk, blindfolds himself, throws the steering wheel out the window and drives towards his sober opponent. The opponent, in full control of his vehicle, has the choice of swerving away and losing or allowing a mutually fatal collision.

The prototype of this sort of political irrationality is the terrorist who appears willing to die in achieving his goals. The hijacker who threatens to destroy both his hostages and himself is believable because his behaviour supports his statements. In a sense, his threats become credible by virtue of the intolerable situation in which he has willingly placed himself. Few governments could emulate this sort of behaviour even if they wished to, however. Rationality is not something which can be deliberately suspended during a crisis and regained for commonplace political and economic negotiations. Hitler's triumph at Munich is often quoted as the classic example of an irrational and headstrong leader gaining victory over a well-meaning but weaker-willed one. However, the Munich settlement was as much a consequence of the Anglo-French belief that Hitler had a reasoned and limited policy towards Germany's territorial boundaries and the question of the Sudetenland as it was due to his threatening and bullying manner. Many believe Stalin to have been clinically insane in his last days but no one doubts that Soviet foreign policy at the time was coherent, reasoned and generally unadventurous. The advantage of irrationality which many writers have found in nuclear diplomacy is a logical consequence of MAD but its practical relevance is doubtful.

The logical conclusion of irrevocable commitment is the infamous 'doomsday' machine which no one seriously wants but which it would be perfectly practical to build. Such a machine might link a massive series of high-yield thermonuclear bombs to a computer which would detonate them automatically under certain stated conditions within its programme. Once activated, the computer could neither be turned off nor its programme altered. Such a machine would be the ultimate in MAD security because mutual destruction would be guaranteed if the machine's conditions were violated. Nor could future leaders be coerced, however timid their disposition; choice would be pre-empted for all time.

The first nation to build this machine would in theory gain the benefit of being able to dictate terms — in effect, a race to pre-empt choice first. Paradoxically, the only decision-maker sufficiently insane to build a doomsday machine probably would not need to bother because his obvious irrationality ought to be sufficiently intimidating in itself. For the computer would be programmed to a number of fixed and inviolable conditions, but no nation could clearly, completely and consistently fix its interests for all time. Circumstances change and policies reflect those changes. Policy requires that commitments be spelled out but it does not demand that they always be explicitly detailed. A full and explicitly detailed statement of what one would or would not do under any imaginable circumstances may just not be believed and may simply let the other know what exactly he may get away with. There are advantages to be gained from political ambiguity, and the fact that certain actions carry the risk of escalating into a major conflict may be a sufficient deterrent in itself.

If it seems possible to draw seemingly inconsistent conclusions from the premises of nuclear diplomacy, it is because the premises themselves are ambiguous and leave room for a variety of conflicting but valid arguments. The acceptance of limited nuclear war or the 'integrated battlefield' — land battles involving conventional, tactical nuclear and chemical weaponry — is consistent with older statements of nuclear politics that emphasized the threat of mutual disaster. Graduated response carries greater credibility than the promise to instigate a mutual catastrophe with no choice other than surrender or suicide; and, should deterrence fail, it is obviously preferable to be able to try and prevent a sequence of mini-disasters from turning into total war. Certainly it is of dubious logic to say that small nuclear wars are basically no different from major ones and that no attempt should be made to distinguish between them nor try to limit a conflict should it occur. This, essentially, is the Pentagon's case for the policy of counterforce and for the development of battlefield nuclear weapons.

Nevertheless, as the Pentagon's critics point out, the belief that the use of nuclear weapons could in some way be limited makes their use more likely. The temptation to start a war would be increased if the attacking power genuinely believed that the outcome would not be an inevitable chain of events culminating in total disaster. Limited wars are competitions not only in weaponry and tactics but also in endurance. The 'winner' would be the one who lasted the longest and, in the context of modern warfare, the one who would be willing to take the greatest risk of continued escalation by carrying on despite the damage already suffered. Limited nuclear war would be the continuation of nuclear politics 'by other means' and all the complications of credibility, determination and uncertainty described earlier would be as relevant as they are to the doctrine of deterrence as a means of preserving the peace. But instead of trying to compete successfully while avoiding war, politicians would be faced with securing peace on favourable terms before the conflict led to complete catastrophe.

World War I began as a conflict of limited objectives for both sides but became fought on total military and political commitment to a final victory almost whatever the cost. On 1 July 1916 the British Army's 60,000 casualties for the first day of the Somme offensive exceeded its losses in the Crimean and Boer Wars combined. By the war's end Britain's casualties on the Western Front alone totalled 2,706,136. This amounted to approximately 6 per cent of the country's 1914 population and nearly 12 per cent of the male population. Yet the vast majority of these losses occurred after the first day of the Somme battle. As happened in World War I the goals and justifications for entering into a war may drastically alter during its course and a 'limited' nuclear war could easily turn into a sequence of devastations which became increasingly indistinct from a catastrophe. It could well be that limited conflict would turn into total war before any of the participants fully realized it or, even if they did, were capable or willing to halt it.

Nuclear Weapons and the Threshold Powers

In September 1979 an elderly American Vela satellite orbiting to monitor the 1963 Test Ban Treaty reported the unique optical 'signature' of a 3 to 4 kiloton nuclear explosion near Prince Edward Island in the southernmost Indian Ocean. Although it has been denied by the parties involved, there is evidence suggesting that the satellite had observed a combined South African and Israeli test of a 155mm extended-range nuclear shell fired from a vessel of a South African task force known to have been in the area at the time. According to a British Independent TV programme, the shell used at Prince Edward Island was a creation of the American firm, Space Research Incorporated — a company with long experience in the design of sophisticated projectiles — and one of a package of 50,000 shells illicitly shipped to South Africa to replenish munitions stocks depleted in the Angolan war.

It has long been widely held that Israel possesses a small stockpile of about 20 Nagasaki-yield nuclear bombs. A highly secret, 2.6 megawatt reactor bought from France — not subject to international inspection — has been operating at Dimona in the Negev since 1963. In addition, there are a number of incidents where nuclear material is suspected of having been diverted to Israel. The most famous is the case of several hundred kilograms of enriched uranium 'missing' from the inventories of the Israeli-linked Numec Corporation of Pennsylvania.

According to a 1976 *Time* report Golda Meir authorized the arming and distribution of Israel's nuclear stockpile during the first hours of the 1973 Yom Kippur war. The exact reasons are unclear but there was at the time a genuine fear that Syrian armour could break through on the Golan Heights and split the country in two.

The Jerusalem-Pretoria nuclear link would be a secret alliance between two politically isolated countries to combine Israeli weapons' expertise with South Africa's abundant supplies of uranium and extensive enrichment facilities. Both countries have the added advantages of well-developed industrial and scientific bases upon which to draw and large reservoirs of unofficial goodwill throughout the Western world and Japan. The CIA, for example, is thought to have illicitly forwarded advanced nuclear weapons technology to Israel in the late 1960s and early 1970s, and South Africa's currently advanced enrichment facilities were partially obtained in a secret agreement with West German industrial and governmental interests.

There are essentially four basic questions involved in looking at nuclear proliferation: the availability of the necessary materials, the sort of weapon that is likely to be made, the capabilities and credibilities of the delivery system, and the political gains of acquiring nuclear arms. It is also necessary to distinguish between nuclear-capable countries which — as far as one can tell — have genuinely renounced nuclear weapons and those which intend to proceed or whose positions are ambiguous. Japan and Canada fall in the first category and Pakistan, Iraq, Taiwan, South Korea into the second. The attitudes of Brazil and Argentina are ambiguous at the moment. West Germany probably has a unique position in being fully capable of developing nuclear weapons and having renounced them but in being likely to cause an immediate East-West confrontation the moment it looked to be reconsidering. India, having carried out 'peaceful' tests in 1974, may or may not have quietly gone ahead with a weapons programme it officially renounces.

The myth still prevails that reactor plutonium is not viable weapons material and that commercial power reactors cannot be used for the illicit production of weapons materials. But there are, in fact, several options open to the would-be exploiter of the ostensibly peaceful reactor, even though such reactors are monitored internationally to prevent their being used to make fuel for weapons. First, certain types of reactor do not require shutdown in order to reload. This may facilitate the clandestine introduction of small amounts of extra material for the illicit production of weapons materials such as plutonium or uranium-233. Uranium-233, which is suitable for incorporation into the simpler type of gun-design bomb, has a plutonium-like fission efficiency and is bred by irradiating thorium-232 with low-energy neutrons. Thorium is plentiful in nature and the two intervening elements decay into uranium-233 over a total half-life of just over 27 days. Like uranium-235 and plutonium-239, uranium-233 is fissionable by slow neutrons and, within a reflector, as little as a 5 kilogram metal sphere may be critical. The degree in which uranium-233 is used in modern weaponry is not known but is thought to be fairly extensive and its military potential was noted from the beginnings of the Manhattan Project. Alternatively, plutonium could be illicitly produced by secretly inserting uranium-238 or natural uranium into a reactor for short periods. It would take a reasonable length of time to obtain a substantial supply of weapon material but, with planning, it would be difficult to detect by international inspectors.

There is also the question of the 'breeder' reactor and its likely spread. Here the reactor is specifically designed to breed fissile material in a 'blanket' surrounding the core in such a way that more of the desired element is created than fuel is consumed. Blanket materials would either be uranium-238 for the breeding of plutonium or thorium for uranium-233. With the spread of breeder reactors there is a growing likelihood that small portions of fertile material may find their way in to illicit hands. Reactor accountancy is not exact; there are always small differences due to error or loss due to natural wastage. Enough minute amounts of diverted material make a viable weapon and carefully 'adjusted' accounting could make the diversions seem within the acceptable limits of material unaccounted for. All reactors fuelled with enriched uranium produce plutonium as a by-product and there have already been numerous suspicious incidents involving unusual discrepancies in plutonium accounts.

The difference between reactor- and weapons-grade plutonium hinges on the reactor's 'burn time'. This is the length of time the rods are subject to neutron irradiation and, because of neutron absorption in plutonium-239, the longer the burn, the greater the proportion of undesirable isotopes. Generally, the shorter the burn, the more 'weapons-grade' the product. Determined countries might either illicitly insert uranium samples in suitable reactors for short burns or arrange maintenance shutdowns to remove rods earlier than scheduled. Again, even with a normal burn, the rods may not necessarily have equal proportion of the higher plutonium isotopes and clever 'stock control' might result in a few with significantly high amounts of 239.

The chief problem facing the clandestine plutonium producer would be to separate it from the unspent uranium and radioactive fission dêbris in the spent rods. When Iraq attempted to buy a 'hot-box' from Italy, a specially constructed laboratory for the remote handling and separation of plutonium, this indicated that she was trying to build an atomic bomb. Reactor sales are normally conditional upon the return of spent fuel rods for reprocessing but an increasing number of countries are either developing their own or are interested in doing so. Pakistan, where the 'Islamic bomb' is being built, is one obvious example and Brazil — which has yet to sign the Non-Proliferation Treaty — is another. A longer-term but more serious danger is the advent of laser enrichment technology.

Unlike uranium, plutonium cannot be significantly enriched by current techniques. A new technique, however, known as laser isotope separation (LIS) is fully capable of breaking plutonium into its separate isotopes via laser irradiation for efficient separation and purification. Thus whatever practical military consideration there is between reactor- and weapons-grade plutonium will disappear for any country acquiring the necessary facilities. LIS will also make the idea of 'denaturing' or 'poisoning' plutonium supplies with artificially high proportions of inefficient isotopes meaningless because they could simply be filtered out. However, even without LIS, denaturing has been criticized as incapable of effectively inhibiting a determined weapon designer — although it would probably make his produce more unpredictable and clumsy.

In the meantime, the chief obstacle to the military use of reactor plutonium is the relatively high proportion of the isotopes 240 and 242 with their high rates of spontaneous fission. This creates a high neutron background and increases the chances of pre-initiation unless the bomb design is capable of assembling a supercritical mass at very high speeds. The higher the assembly speed, the lower the probability of efficiency-destroying premature fission. Implosion designs capable of assembling high-density supercritical assemblies at speeds in excess of 2 to 3 kilometres a second reduce the chances of pre-initiation to minimal proportions and, even at these speeds, have high probabilities of performing as the designer intended. The implosion-profiling weapon described on page 14 is easily capable of assembly speeds well in excess of the necessary velocity and of compressing the core to four or five times its normal density. The system can also be adapted to a cylindrical system which is in some ways simpler than the spherical although it will be less efficient. The designer even gains the extra benefit of avoiding the added complexities of incorporating a neutron source. With rapid assembly, the neutron background of reactor plutonium will act as an effective fission initiator as the core reaches and passes critical density rather than as a barrier to efficient weapon performance, and many of the superpowers' modern nuclear weapons are thought deliberately to incorporate a high plutonium-240/242 content as the source of fission initiation.

The same considerations apply to amateur bomb builders with adequate facilities and equipment and sufficient expertise. The amateur's efforts may be additionally rewarded in that even poor weapon performance may be highly effective for him whereas would-be nuclear powers require predictable and effective engineering.

As yet, no candidate nuclear power appears to have opted solely for the reactor plutonium short-cut. Instead they have tended generally to repeat the efforts of the Manhattan Project and have invested in uranium-enrichment facilities and the production of predominantly weapons-grade plutonium. The benefit is effective weapons with simpler 'Trinity' designs but with the constraint of having to base them on weapons-grade materials. The difficulties of secret testing compound the difficulties; the more sophisticated the design and the less the prior experience, the more the need for confidence-building tests of nuclear as well as non-nuclear components. This is true of even Trinity-level weapons. Despite kiloton-range yields, for example, India's 1972 tests reportedly fell far short of expectations.

Even smaller versions of a weapon like the Nagasaki bomb would require aircraft delivery. Although any of the modern fighter-bombs in service with the threshold nuclear powers would be effective as delivery systems, they would not necessarily be credible ones. Modern air defences are making the fighter-bomber increasingly vulnerable and, assuming that an emerging nuclear power wishes to threaten a neighbour with unacceptable damage, the deterrent may lack credibility because it could not be guaranteed to penetrate air defences. If the opponent is technologically advanced, populous and geographically dispersed the problem is even more acute. Neither South Korea nor Taiwan — who are widely suspected of having already secretly assembled nuclear weapons — would be likely successfully to penetrate China so as to cause anything approaching even the bare minimum of damage associated with superpower deterrence. India is in the same position *vis-à-vis* China but not, as yet, as far as its immediate opponent and chief nuclear concern, Pakistan, is concerned.

Emerging nuclear powers who desire a scaled-down version of nuclear deterrence will eventually be forced into developing effective delivery systems. So far only Israel has made the effort. In the mid-1970s she was known to be developing a 450 kilometre (280 miles) surface-to-surface 'Jericho' missile capable of carrying a nuclear warhead. A longer-range, 1600 kilometre version — bringing most major Arab cities within targeting distances — was under development at the same time. Jericho was known to be suffering guidance problems and little or nothing is known about its current status. A small but secret production, however, is entirely possible and the announcement only deferred in light of negotiated peace prospects. India is the only other threshold nuclear power with the immediate prospect of a medium-range ballistic missile. She is currently working on an advanced version of its SLV-3, home-built space rocket capable of carrying a 150 kilogram payload into orbit. The up-dated SLV-3, however, would not threaten China without extensive improvements to range and payload. South Africa has its own well-developed short-range military missile programme but probably has access to Israel's Jericho technology as well.

However, the possession of even a few unwieldy nuclear weapons is strategically valuable whatever the credibility of the delivery system. First of all, the chance that even one or two might be successfully delivered is a risk that no rational power would undertake in anything but the most extreme circumstances. Secondly, the possession of nuclear weapons exercises a disproportionate political influence on the superpowers. Israel's weapons ensure that neither Washington nor Moscow can allow Israel to be threatened to a point where a precedent-shattering resort to the nuclear option appears likely.

Taiwan, South Korea and South Africa gain the same advantage from a nuclear capability; American cooperation and Soviet hesitation follow from the wider dangers of confrontation politics. At this level, differences between battlefield and retaliatory uses of even one weapon are much more marginal than they would be otherwise and the political utility of any nuclear capability works for any aligned or unaligned country; the advantage follows from public uncertainty and the 'unofficial' knowledge of the world's governments that a capability actually exists. The 'status' gained as a member of the nuclear club is another consideration, but, so far, has only arisen in India's admission of its 1974 test, Libya's futile efforts to buy a weapon from China some years ago, and recent references to the emerging Islamic bomb. By contrast, countries committed to developing nuclear weapons — including Iraq, Pakistan, Israel and South Africa — have made every effort to deny publicly their intent while blatantly emphasizing their capability.

The eventual proliferation of hydrogen weapons is infinitely more difficult to predict. Together, Israel and South Africa already possess the expertise and rudimentary industrial base required. They have the nuclear technology to provide triggers for militarily feasible and compact hydrogen bombs. But to go ahead would require considerable finance, research and expansion of nuclear and related industries. Considerable testing of at least the nuclear trigger components would be essential and eventually a full-scale trial run would be necessary. It would be virtually impossible to keep such a programme secret and powers turning a blind-eye to clandestine nuclear weapons programmes might well react quite differently if faced with H-bomb proliferation. An Indian and perhaps a Taiwanese H-bomb could follow an Israeli-South African one within a few years and Pakistan could conceivably manage a programme with its more meagre resources. But even countries willing to make the financial and industrial sacrifices would have to ask themselves if success would gain them any obvious military or political advantage not largely present in their existing nuclear programmes.

6 Anti-Missile Defences and Beam Weapons

Anti-Ballistic Missiles: Another Chance?

The anti-ballistic missile (ABM) system under consideration for MX is, with improved technology, essentially the same as one developed over a decade ago and negotiated away in the 1972 ABM treaty. Signed around the time of SALT I, this treaty initially limited both America and the Soviet Union to 200 ABMs, a figure which was later modified to 100. The ABM treaty is reviewed every five years and there is an increasing belief in Washington that the United States should renegotiate the terms in order to take advantage of recent technological advances — advances which appear to be far ahead of Soviet work.

As well as defending MX sites, the new ABMs could be placed on board ships or land vehicles to protect armies in the field or American bases abroad. In theory ABM protection could be given to all American and allied cities but, as yet, this does not seem to have been seriously considered as blanket anti-missile defence is politically destabilizing. It removes the hostage relationship upon which MAD ultimately rests and thus affects the whole basis of deterrent strategy.

Before the 1972 treaty, the United States considered two ABM schemes, known as Sentinel and Safeguard and based upon two missiles, Spartan and Sprint. The schemes differed principally on the degree of protection they offered to the American population and to Minuteman sites.

The Spartan missile was designed to intercept incoming warheads outside the atmosphere at distances up to several hundred kilometres. It carried a single megaton-range warhead which would destroy an attacking re-entry vehicle through neutron and gamma radiation (very short wavelength X-rays) which easily penetrate matter. Sprint, on the other hand, was smaller with faster acceleration, able to climb a mile within a few seconds. It was intended to provide a second layer of defence by intercepting warheads which had bypassed Spartan as they re-entered the atmosphere in the last stages of their flight. Because it was designed to detonate at altitudes as low as 8 kilometres (5 miles), to minimize possible surface damage from overpressure, heat, EMP and radiation, Sprint was equipped with a low-yield warhead of a few kilotons designed to destroy chiefly through neutron radiation. It was the technological ancestor of the neutron 'bomb'.

Each ABM site was to have utilized two distinct types of specialized radar: a long-range or perimeter acquisition radar (PAR) and a missile site radar (MSR). The purpose of PAR was to detect and track an attacking missile several thousand kilometres away and, in conjunction with a sophisticated time-sharing computer network, compute its trajectory and final impact point. The computer would calculate the relative threat from a host of incoming signals and hand the target over to the MSR for continued tracking. With a computed intercept point, an ABM would be launched by computer command and rise to meet the incoming warheads.

The use of penetration aids makes the task more difficult. In addition to live warheads, ballistic missiles carry a range of devices designed to fool defensive radars. These include the release of metallic strips or 'chaff' which reflect radar signals and cause hundreds of targets to appear on the screen. Metallic-coated balloons may be released along the trajectory and imitate more of the characteristics of actual warheads than chaff (which is fairly easy to distinguish) until the actual re-entry stage is reached. In addition, débris from the second-stage missile booster can be made to follow the warhead to further fool defensive radars. At the point the warhead begins to re-enter the atmosphere, chaff, balloons, and débris are easily distinguishable from the real thing but there are so-called 'heavy' penetration aids which can be used as well. These are dummy warheads that carry packages of electronic spoofing and jamming devices which follow trajectories designed to imitate live warheads through re-entry.

For ABM intercepts, the problem would have been to determine the correct targets from the spoofing materials, since too many mistakes would rapidly deplete the available ABMs. One obvious way is to ignore 'warheads' with trajectories that lead away from targets of any importance but this can be countered by designing attacking warheads which manoeuvre during flight. The warhead is sent on a false trajectory but corrects itself at the last minute and flies to the intended target. It is thus vital to differentiate between the shape, speed and weight of the various signals since live warheads have significantly different radar signatures from those of spoofing devices. Here, one of the most crucial advances in radar technology since the 1972 ABM treaty is the use of image-building high-to-low frequency radars. A laser-based system known as LADAR is nearing procurement. Used together, these systems are capable of building up such a detailed image that a genuine warhead can be identified from the most sophisticated dummy with virtual certainty. Improvements in ABM guidance and accuracy now enable a missile to kill incoming warheads with conventional rather than nuclear explosives.

Target discrimination in any new ABM will depend on an integrated system of radars and infra-red sensors distributed on the ground, in satellites at altitudes of 45,000 kilometres (28,000 miles) and aboard the killer missile itself. The US Army is testing focal-plane mosaic long-wave infra-red sensors in missiles and an on-board computer capable of handling the massive amount of data generated by the wide view offered by these mosaic detectors. Each sensor will contain some 10,000 separate micro-elements. LADAR, a pulsed laser detector system, provides extremely sharp definitions of the target with its very narrow beam. With pulsed LADAR, it is possible to get a signal return from dust particles and even air molecules in the upper altitudes. In order to reduce the possible influence of electronic jamming devices, small radar phase-shifters based upon solid-state circuits are under development for millimetre-length radars in the higher frequencies.

To destroy an incoming warhead without using nuclear explosives, the ABM must be able to draw to within a few metres and be capable of following any evasive manoeuvres. The defending missile will track the incoming warhead with its own package of infra-red sensors and radar and home in for the kill by reading changes of direction or speed through detailed analysis of Doppler shifts by the on-board computer and by information relayed from the ground.

Any new ABM deployment will be two-tiered and have some missiles capable of intercepts in space and others for destroying intruders in the upper atmosphere. In the atmosphere a short-range ABM will kill with a warhead of conventional explosives. However, blast is ineffective in the vacuum of space, so long-range ABMs will use a different method — lasers, for example, or bombardment of the re-entry vehicle with shrapnel or hundreds of metal pellets. The reaction times of the new ABMs have been increased. The short-range missile, for example, will accelerate to a speed of 9325 kilometres an hour (5800 miles an hour) within 1.5 seconds and reach an altitude of 1500 metres (about 5000 feet) in about 1 second. It will detect and begin to track the incoming re-entry vehicle at about 15,000 metres (50,000 feet). It may soon be possible to extend the effective range of ABMs to Soviet airspace in order to kill missiles even before they leave the atmosphere. Such a system would be linked to a network of early warning satellites, long-range and rapidly acting missiles and ground-based radars capable of receiving signals from over the horizon. According to the Pentagon, the necessary technology will become available before the middle 1980s.

There are many objections to ABMs, their cost for one, an ABM system being estimated at upwards of $100 billion. Another is the relative ease of countering the previous ABM system with improved penetration aids, added warheads and manoeuvrable re-entry vehicles which was an important reason for the willingness to scrap it in 1972. But the ABMs of the 1980s and 1990s promise sufficient advances to make these countermeasures much less effective. Defending ICBMs alone would be less provocative than a nation-wide ABM coverage of both military and civilian targets. A limited short-range coverage of strategic missiles would leave enough urban areas 'open' to give the other side a reasoned belief that he could still hold his opponent's population as a 'hostage' to good behaviour. However, long-range ABM systems — especially those designed to intercept missiles in Soviet airspace — would be seen as offensive in intent, threatening all missiles launched from the Soviet Union and not just those homing in on American ICBMs. Such systems would, at best, be dangerously ambiguous.

Laser Weapons

Almost from the moment the first practical lasers appeared in 1960 their 'death ray' potential was dwelt upon in lurid detail, but their first military applications were in the more mundane areas of range-finding and sighting techniques. Here, the laser's ability to make precise measurements made it ideal. Using lasers as actual weapons, on the other hand, has taken much longer to achieve because their suitability has in the past been in serious dispute. However, recent technology has opened up new possibilities.

Until 1980, the United States concentrated upon developing lasers as surface-to-air defences for warships and other high-value targets; since then, however, fearful of Soviet progress, the Pentagon has concentrated on developing orbital laser platforms to protect reconnaissance and other strategic satellites and as a defence against ballistic missiles. At the same time, the possibility of using tightly aimed beams of accelerated sub-atomic particles against space vehicles and ballistic missiles now appears to be sufficiently plausible, after years of controversy, to warrant expanded research programmes.

'Laser' is an acronym for light amplification by stimulated radiation. It refers to a technique for producing a parallel and coherent beam of monochromatic light, one consisting of a single wavelength or colour. The beam is created by the stimulation of a gaseous, liquid or solid medium between two aligned mirrors. Atoms exist at various discrete levels of energy. By adding another discrete amount of energy, the atom can be stimulated to a higher state. To raise the electron of a hydrogen atom from its lowest or ground state, for example, requires 10.2 electron volts and a further 12.1 eV to take it to its second level of excitation. In the laser the atoms are 'engineered' to release their radiation in phase with the stimulating wave, which has an amplifying effect and produces a monochromatic beam. By passing the wave back and forth between the mirrors before it exits, the energy level can be amplified even further. When the radiation's wavelength is in the visible part of the spectrum the device is known as a laser (for 'light amplification'), but when it is in the microwave category it is called a maser (for 'microwave amplification'). Lasers and masers (which, for simplicity, may be thought of as the same thing) can be 'pulsed' by breaking the beam via an opaque shutter which, upon firing, is rendered clear by an impulse of electric current that turns the stored energy into a pulse of radiation. Even more powerful energies can be created with oscillating lasers where the beam ranges over a number of synchronized frequencies to produce shorter and more intense pulses. Certain types, such as the 'free-electron' laser, are tunable and may be adjusted to work at either end of the spectrum or in the visible ranges.

The American Defense Department has recently chosen three laser types for weapons application: these are the hydrogen fluoride, argon or krypton fluoride, and the free-electron. The first functions towards the infra-red end of the spectrum and the second towards the ultraviolet while the last ranges between both. According to *Aviation Week and Space Technology* the free-electron laser was developed at Los Alamos in New Mexico and works by converting electron kinetic energy to optical radiation by interacting an electron beam with a periodic magnetic field known as a 'wiggler'. Electrons are forced into the accelerator and then into the magnetic field where they lose energy to the photon field as they leave. A portion of the beam's energy is recoverable and is used to accelerate the next collection of electrons.

The Lawrence Radiation Laboratory in Livermore, California, is concentrating upon developing argon or krypton fluoride lasers as anti-missile systems.

These two types of pulsed laser are the most efficient of the ultraviolet group but require large amounts of power. For this reason, miniature thermonuclear explosions have been posited as a likely source of power. One approach currently under study is to power the laser by bombarding microgram amounts of deuterium and tritium with laser light or charged particle beams to induce small fusion explosions. In a satellite system such a power source would supply several lasers.

Generally, the shorter the wavelength, the more suitable the laser due to the possibility of smaller optics. The krypton fluoride excimer laser has a wavelength of roughly 0.248 microns compared to 2.7 microns for the hydrogen fluoride laser. This is the shortest wavelength of the major candidates. The free electron tunable laser may operate over the range of 0.3 to 0.5 microns. This flexibility gives it the additional advantage of being more difficult to jam. Long-range anti-missile lasers could require 20 to 60 megawatts of power and mirror diameters of 25 or 30 metres. Shorter-range or site-based anti-missile lasers, on the other hand, could require as little as 5 megawatts of power and mirror diameters or only 4 or 5 metres.

These lasers would destroy a ballistic missile by punching a hole through its surface and causing it to disintegrate in mid-flight, or at least weakening its structure and vital components. The chances of a kill are increased if the missile is hit while it is accelerating, and the idea of basing lasers in space is to hit the missile during its launch phase as it accelerates out of the atmosphere. There are reports of a 1980 Pentagon estimate that 25 orbiting 25-megawatt chemical laser stations could kill as many as 1000 ballistic missiles in the atmosphere. But the demands upon optics and control systems are immense. A Los Alamos study concluded in 1980 that the required technology was beyond current capabilities, although not impossible. A continuous laser study funded by the Defense Advanced Research Projects Agency will attempt to construct and ground-test, by 1983, a 1.2-metre (4-foot) optics system suitable for space-based lasers.

One of the first military projects scheduled for the Space Shuttle programme is a series of test runs for a control, aiming and tracking package suitable for space-based lasers. Data correlation, beam accuracy and stabilization are critical, and the Los Alamos study concluded that an angular precision of less than one-millionth of a radian was necessary if a ballistic missile was to be hit at ranges of 4830 to 9650 kilometres (3000 to 6000 miles). The Space Shuttle test programme — code-named Talon Gold — will attempt to hold laser accuracy to within about 20 per cent of the required millionth of a radian.

The Soviets have already used their Cosmos satellites and Salyut space station to test laser techniques and the advantage they are presumed to hold over the United States in these weapons was one of the chief reasons why Defense Secretary Harold Brown announced, in 1980, a switch of emphasis in laser research from short-range, ground-based applications to the less certain, but potentially more important, possibilities of space-basing. This decision was made contrary to the recommendation of a Defense Science Board brains-trust that had been commissioned to investigate the most fruitful lines of Pentagon-funded laser research. Nor was the decision welcomed in all sections of the American armed forces; the Navy, for example, was especially keen on the rapid development of lasers for ship-board defence systems.

Far more immediate than employment in anti-missile systems is the use of lasers against satellites. Disintegrating the satellite in science-fiction style is not the aim; instead, the objective is to 'blind' it by damaging its sensitive electronics and optics, as laser light is capable of damaging multilayer insulating lens coatings or weakening the glue holding them in place. A power of 100 megawatts or even less per square centimetre may destroy the glass itself. The Soviet Union probably already possesses a carbon-dioxide pumped laser capable of attacking satellites in low altitude orbit. This would threaten America's Big Bird and KH-11 spy satellites which fly over the Soviet Union at altitudes of approximately 210 kilometres (130 miles). More doubtful are reports of Soviet progress in short wavelength pulsed iodine and excimer lasers capable of attacking satellites at altitudes above 4825 kilometres (3000 miles).

Iodine, excimer and free-electron lasers are thought to be potentially more efficient within the atmosphere and it is believed that the Soviets are concentrating on ground-based systems for the moment. Intelligence sources claim to have identified a Soviet effort to develop these laser types at Krasnaya Pahkra near Moscow and a short-range laser air-defence system at Golovnino. But other reports talk of a 114,000-kilogram (250,000-pound) 12-man space station and a 5.4 million-kilogram (12 million-pound) thrust booster for taking directed-energy weapons into space. Although many experts argue that neither side will overcome the engineering difficulties surrounding space-based lasers for at least the next five years, others insist that a Soviet breakthrough is imminent.

Another Soviet programme increasingly reported is the development of a ground-based laser capable of attacking satellites at distances of up to 40,000 kilometres (25,000 miles). This would bring within range America's three early-warning satellites fixed in geostationary orbits at altitudes of 35,400 kilometres (22,000 miles). Lasers thus have three potential military roles: as short-range air-defence systems, long-range satellite killers based either on the ground or in space, and as space-based, long-range anti-missile defences.

Anti-satellite warfare is not restricted to laser development. Satellites are vital to the superpowers' military machines because reconnaissance, navigation and communications depend upon them, and something like 80 per cent of America's defence communications are dependent upon satellite relay. In a war satellites would be even more important as conventional communications became disrupted by the after-effects of nuclear explosions. As a result of their importance the United States is experimenting with small satellite-killing missiles which could be fired from high-flying jet aircraft. Also, both America and the Soviet Union have researched killer satellites which, after a series of manoeuvres, draw alongside their targets and detonate, destroying them through such mechanisms as a shrapnel shower. The American Space Shuttle is known to worry the Russians for its anti-satellite possibilities, as it would be capable of drawing alongside a satellite and either destroying it, damaging it or removing it from orbit and carrying it back to earth for examination, and eventually might be equipped with satellite-killing lasers or missiles. Workable lasers are attractive as satellite killers because they would be less expensive, quicker and easier to employ than using other satellites or space vehicles.

Air defence lasers based aboard warships, land vehicles or — as in a reported case of Russian research — railway cars are close to being introduced. In 1980 the Americans revealed a test aircraft carrying a prototype air defence and possible anti-missile laser. The plane was the ASAF-KV-135, a modified Boeing-707, which has been designed to carry several tons of laser equipment, fuel and controls. It is now undergoing a long series of test intercepts of target drones and air-launched guided missiles and the first, in January 1981, was successful. Space-based anti-missile systems are less imminent, both because of the demands placed upon sophisticated engineering and an uncertainty surrounding their overall effectiveness. Laser counters for ballistic missiles might, however, be fairly easy and economical to introduce by equipping the missiles with shielding or reflectors, for example, or by causing them to spin during flight to prevent beam concentration.

Particle Beams

Since 1977 the former head of US Air Force Intelligence, Major-General George Keegan, has been something of a voice in the wilderness. General Keegan and his staff had for years been warning that the Russians were developing weapons utilizing beams of accelerated sub-atomic particles and that they would begin to deploy them towards the middle of the 1980s. Few authorities paid attention and Keegan resigned his office to become a vocal critic of American policy. The dispute began over the interpretation of intelligence information, including satellite data, concerning the presence of gaseous hydrogen with traces of tritium in the upper atmosphere. These findings, dating back to 1975, were said by the US Air Force to relate to particle beam work at a Soviet facility in Semipalatinsk which was researching contained thermonuclear explosions as a pulsed power source.

The feasibility of beam weapons for the near future has been hotly disputed. In 1979 the Massachusetts Institute of Technology published a study concluding that, for the moment, beam weapons presented operational and technical difficulties which appeared insurmountable. And two years previously Defense Secretary Brown told the National Press Club in Washington that solving the problems involved was impossible.

Unlike lasers, which use photons of electro-magnetic radiation, beam weapons would accelerate and focus a beam of sub-atomic particles which have mass and do not reach the speed of light. Neutral particles — including electrons or protons — would bombard the target and effectively destroy it through the destructive nuclear reactions they would cause. All ions produce braking radiation when they decelerate in matter but the smaller their mass and the higher their velocity, the greater the effect. The electron is the smallest charged particle known (both the proton and the neutron have over 1800 times the mass) and can reach near-light speeds. At such velocities, electrons will produce extremely damaging braking gamma radiation as well as X-rays. In effect, an electron beam weapon would be a beta 'gun' and a portion of the gamma rays produced in the target will be far more energetic than those occurring naturally.

Recent events have caused considerable consternation that General Keegan may have been right. This shift is largely due to a Soviet construction at Sary-Shagan near the Chinese border. Code-named Tora, the Sary-Shagan project was begun in November 1979 and has been under observation by American satellites. What disturbs some American defence observers is the suggestion that Tora is a prototype beam weapon based upon betatron-accelerated electrons. However, others believe it is a pulsed iodine laser facility or the beginnings of an experimental fusion power station.

There is little doubt that Russian work in this entire field equals that of the West and may even be further ahead. Project White Horse, one of America's most advanced particle beam programmes at Los Alamos, uses a device known as a 'radio frequency quadrupole' (RFQ) which accelerates and focuses particles into coherent beams using electrical fields only. The RFQ allows particle focusing independent of velocity, unlike earlier devices which use magnetic fields, and also incurs only minimal particle loss as it gathers in and accelerates even those with low velocities. The overall result is less degradation of the beam produced. The RFQ is a result of Soviet research and was developed at Los Alamos after a careful reading of data published in Russian scientific journals. It has a wide application in such areas as fusion energy development and as a beam weapons base.

In 1980 *Aviation Week and Space Technology* published details of the Sary-Shagan site which it said were gleaned from intelligence and military sources. The site is said to consist of 12 magneto-explosive generators behind shielded walls. The generators, said to be portable, are the result of work begun by the Soviet physicist and dissident Andrei Sakharov and involve the conversion of chemical explosions into the amounts of electromagnetic energy required to power (around 1000 MeV per pulse) a tight beam of accelerated electrons. They do this by injecting a high magnetic field into a tiny cylindrical space and then compressing it by imploding a surrounding metallic liner. The resulting electrical energy is relayed by power lines into an electron accelerator described as a linear air-cored betatron. The betatron accelerates the electrons and focuses them into a coherent beam which then leaves through an aimable nozzle at almost the speed of light. In effect, the system would amount to a beta radiation 'gun' as beta rays consist of high-velocity electrons.

The Soviet Tora programme is thought to be principally the work of physicist A. Pavlovski, who has researched betatron accelerators and pulsed-power magneto-explosive generators for years; even American generators of this type are commonly called Pavlovski generators. It was due to be completed in the summer of 1981 and, if the weapon theory is correct, testing on airborne targets would have begun shortly thereafter.

Tora is more likely to be a prototype of a ground-based weapon rather than one destined for an orbital platform since an electron beam could not be effectively used outside the atmosphere. Electrons, being negatively charged, would repel each other in the vacuum of space causing the beam to spread and weaken. But within the atmosphere, such a beam is thought to create magnetic conditions which counter this natural repulsion and hold the electrons together in a tight pattern. Space-based beam weapons would use neutral particles, whose lack of electrical charge would produce a narrow beam.

America's hopes of a neutral-beam weapon currently rest on the White Horse programme at Los Alamos and, for a ground-based system, upon an electron beam accelerator under development at the Lawrence Radiation Laboratory in Livermore, California. White Horse seeks to accelerate a negatively charged beam of hydrogen, neutralize it and then focus it at targets up to thousands of kilometres away. A least a decade's work is thought to be involved but, if successful, it would eventually lead to the placement in space of rings of satellites housing accelerators, tracking and sensor equipment for a neutral-beam defence of the United States. The Livermore project involves the exploitation of chemical or nuclear explosions as a power source for electron acceleration. An advanced accelerator is due for completion in 1982. It is designed for the rapid emission of high-energy pulses of around 50 MeV and is expected to demonstrate the ultimate feasibility of ground-based charged-particle weapons once and for all.

Doubts about harnessing the power necessary for such a beam weapon and its focusing and control within the atmosphere have been widely expressed. The earth's magnetic field, for example, may deflect an electron more than 200 kilometres (125 miles) in a shot of 5000 kilometres (3100 miles). In addition, using such a weapon under wartime conditions would be far different from peacetime experimentation. Nuclear weapons effects such as electromagnetic pulses could disrupt electron beam propagation and a number of explosions in the vicinity of use might make the weapon virtually unusable for a considerable time. Space-based beam or laser anti-missile weapons would also pose the same threat to deterrent strategy as ABMs. Doubts on the nature of Tora or upon the military feasibility of beam devices have far from disappeared; they have simply lessened in the face of increasing evidence of their ultimate plausibility.

Railguns and Mini-ICBMs

An old idea that might protect the ICBM from advanced anti-missile defence is the railgun, which has already proved capable of accelerating 12-millimetre (½-inch) projectiles to speeds of 10 kilometres per second (22,370 miles per hour). This is roughly 10 times the speed of a rifle bullet and in line with a spacecraft. In theory, speeds of 150 kilometres per second (335,556 miles an hour) are obtainable.

The possible uses for railguns range from a fusion energy source which bombards deuterium-tritium mixtures with super-velocity particles to guns capable of piercing the toughest armour with ease. Equally, advanced railguns could cold-launch ballistic missiles or spacecraft by hurling them beyond the atmosphere before their engines ignited. Because detection by satellite depends largely on the heat generated at ignition, missiles might not be detected until well after launch; missiles could also be made much smaller as they would no longer require their launch stages to escape the atmosphere. Adopted into an anti-ballistic missile system, railguns could destroy incoming warheads. If based aboard satellites or spacecraft they could be used to destroy missiles as they left the atmosphere by hitting them with projectiles travelling at extremely high velocity. The projectiles need not even be explosive because, at the levels of velocity envisaged, momentum alone would be sufficient.

According to Livermore scientists, the Russians had performed about 150 railgun tests by late 1980 as against 40-odd by the United States. Like much of the work with particle beams and high-energy lasers, American research on railguns has been jointly financed by the Defense Advanced Research Projects Agency and the Department of Energy.

The theory of the railgun goes back to the writings of Jules Verne where it was first proposed as an infinitely more efficient means of propelling a projectile than gunpowder. Scientists have experimented with the idea for many years and the Germans investigated its military possibilities during World War II but sustained work revived in the West only a few years ago at the Australian National University in Canberra.

A current American experimental railgun consists of two parallel copper rails with a plastic projectile between them. When the gun is fired, a current applied to one rail passes through a metallic film fuse at the base of the projectile. It instantly vaporizes it, creating a plasma, and passes to the other rail to complete a circuit and produce a power of around 1 million amps. The current in the plasma creates an electromagnetic field which reacts against the one in the rails. The electromagnetic field is forced forward, accelerating the projectile in front of it. The principle is the same as an electric motor but, instead of imparting a spin to a rotary wheel, the magnetic field forces the plasma and projectile outwards in a straight line guided by the rails.

To increase acceleration one rail is coated with a strip of conventional explosive. As the gun is fired, the explosive is ignited at the rear of the rail where it forces the imploded rail into contact with the other. The explosion proceeds from the rear of the railgun to the front. As the two rails come into contact along an increasing proportion of their length, the magnetic field is compressed into an ever smaller space which further accelerates the plasma and the projectile. The firing sequence takes around two-thousandths of a second and when the projectile leaves the gun it does so at maximum velocity. At the moment, the projectiles used are small plastic cubes that measure about 12 millimetres (½ inch) on each side but non-conductive metal missiles are also being researched.

Practical railguns, military or civilian, are a long way from reality because of a large number of technical obstacles. At present the gun is a single-shot device only, since the explosion and high projectile velocity effectively destroy it with each firing by tearing away insulating material and metal. In addition, the velocity of the projectile is so great that it may tear apart or burn up as it meets atmospheric resistance upon leaving the gun. Scientists believe that these difficulties can be solved, but any weapons development would require very much higher standards of durability and reliability than peaceful applications because a gun which required a long reloading time or which became unusable after even a few firings would be considered impractical for deployment. Although space-basing would solve most of the problems associated with using the projectile within the atmosphere, durability and the automatic reloading of the gun would be even more essential.

However, the principles of the railgun have dramatic potential for nuclear missiles. Future developments in the miniaturization and sophistication of electronic guidance and memory systems combined with a significant reduction in warhead size could lead to the production of a generation of small ICBMs with advanced capabilities but individually reduced to the dimensions of today's tactical missiles. Exploitation of the higher transuranic elements could feasibly reduce warhead size dramatically while current work on such things as holographic and bubble memory already promise further quantum jumps in the sophistication of electronic circuitry. If and when the railgun is developed into a viable weapon, it would be possible to use it to launch miniature ICBMs with only sufficient on-board guidance and power to see them through re-entry and a terminal path to the target. Launch sites could be easily concealed and the missiles could be produced economically in the thousands. The prospect is one of an intercontinental artillery which could overwhelm missile defences by sheer numbers and by drastically reducing the warning times. The necessary technology is still years away but, if it materializes, there may be a proliferation of cheap, mobile and sophisticated ICBMs in such numbers as to make any arms limitation agreement futile.

7 Radiation Weapons

The Neutron Bomb

In April 1978 unofficial messages on Pentagon bulletin boards protested against Carter's postponement of the neutron bomb. They read: 'Bows and arrows kill people but leave buildings intact.' In August 1981 the controversy broke anew when Reagan announced that neutron warheads would be produced and stockpiled in the United States but, for the time being, not within Western Europe.

Probably no single weapon has caused so much furore and confusion. Partly as a result of a skilful Soviet propaganda campaign it has become *par excellence* the property-preserving, capitalist weapon. (However, when France reported a successful neutron bomb test at Mururoa atoll in June 1980, the event passed virtually without comment.) But because of its low yield it is only intended as a battlefield weapon and would have little value in an all-out nuclear attack on cities.

The neutron bomb is also known as an enhanced radiation/reduced blast (ERRB) nuclear weapon and is designed to kill chiefly through neutron and gamma radiation rather than heat and blast. Work began around 1958 and the first successful test of such a device was carried out in 1962 for an ABM warhead. Later the technique was seen as useful for attacking concentrations of armour and a partial answer to the Warsaw Pact's decisive superiority in tank numbers. Tanks have a high degree of survivability against the blast and heat of tactical nuclear weapons unless they happen to be extremely close to the point of detonation, but their crews can be attacked by killing or incapacitating radiation at longer ranges. Furthermore, any relatively large-scale use of the older tactical nuclear weapons on a European battlefield would soon devastate the entire area with fallout, killing thousands and perhaps millions of civilians and allies downwind. In this regard, another justification for the neutron bomb was the emerging doctrine of flexible response, which required an ability to be able to minimize collateral damage even within enemy territory without unduly sacrificing weapon effectiveness.

On both grounds the enhanced radiation/reduced blast technique became an attractive proposition. Such a weapon could kill through radiation within a maximum radius of some 1500 metres, blast damage would be greatly reduced and radioactive contamination would dissipate within a period of a few minutes to an hour or so. Its potential for killing an urban population without destroying the city was widely publicized, but the weapon was no more designed for this use than many of the other weapons and techniques developed for flexible deterrence.

The makeup of a neutron weapon is not openly known; what is known is that it is a small fusion (hydrogen) bomb with a yield of around 1 kiloton. But there are important differences between the ERRB and the sort of hydrogen weapon described in Chapter One.

To begin with, neutron weapons are 'clean' and have no fallout producing uranium-238 in their driver mechanisms. The nuclear proportion has been reduced as far as possible and only constitutes a small percentage of the overall yield. In a truly 'clean' weapon, the driver would be made of dense materials which give off minimal radiation under neutron bombardment. But since the purpose is to maximize neutron production the fuel probably has an outer profiling layer composed of an element such as beryllium, which gives off neutrons when struck by either the helium-4 nuclei (alpha particles) or the neutrons produced in the fusion process. In certain cases, an element like beryllium will give up two neutrons when struck by one.

The beryllium layer would thus act as a neutron multiplier. To take a simple example, the chief reaction in a neutron weapon is the tritium-tritium fusion which results in one nucleus of helium-4 and two neutrons. On a yield-for-yield basis, this reaction will produce two to three times the neutrons of the other important fusion reactions listed in note 2, making tritium the basis of a radiation-enhanced fuel. Should each of the three particles produced then react ideally with beryllium, the end result would be five neutrons radiating outwards — a 2.5 neutron increase on top of the already radiation-enhanced fuel. In addition, since deuterium and tritium ions will produce neutrons in beryllium, the profiling layer would act as a neutron source for the unused fusion débris. It is a common mistake to think of the neutron bomb as a radically new 'death-ray' weapon; its essential secret is the tailoring of specific materials and design to create maximal radiation with minimal blast and fallout.

Neutron weapons are only one example of clean hydrogen weapons and their 'opposite', a blast-enhanced/radiation-reduced warhead, has reportedly been developed. It would probably work by maximizing the number of high-energy yielding, non-neutron producing fusion reactions with a specific fuel mix. Such a weapon might be used to attack specific targets within urban areas without the radiation hazards associated with less tailored-effect clean warheads. Like neutron weapons, radiation-reduced warheads would have small yields to minimize unwanted collateral blast damage but it would also be necessary to fit them to highly accurate delivery systems — both to insure that the target is destroyed and to avoid unintended devastation through near misses.

The basic fuel in neutron weapons is almost certainly lithium-6 tritide. This gives a greatly enhanced neutron output but tritium fusions are comparatively difficult to ignite. Triggering has to be designed to implode the fuel to the higher densities and temperatures necessary for an effective burn. Some deuterium may also be included in the fuel to first ignite and accelerate the process by its reaction with tritium. The deuterium may be mixed into the fuel (making it deuterated lithium tritide) or may be included as a separate layer of lithium deuteride.

Lithium tritide weapons would have relatively short shelf-lives due to tritium's radioactive half-life of only 12.3 years. Because all nuclear weapons are brought in for frequent overhauls, checkouts and refittings, this would not be a severe handicap since the fusion cylinders could be periodically replaced or up-graded. But Reagan's decision does mean an increased demand for tritium. This creates problems since America's principal tritium suppliers — government reactors at Hanford, Washington and Savannah River, Georgia — also happen to be the chief sources of weapons-grade plutonium.

Even without refinements, a 1-kiloton hydrogen weapon would yield copious quantities of neutrons and gamma rays. Of the two, neutrons are more lethal but as the distance from the explosion grows the proportion of gamma rays to neutron radiation increases. The reason for this is that the neutrons are steadily captured by the atoms they encounter which results in an exponential decline in the neutron population over distance. Neutron capture frequently causes the newly excited atom to emit gamma radiation before it returns to its ground state. Because gamma rays travel outwards in a straight line, they are more easily protected against than neutrons, which may come from all directions after colliding with an atom and recoiling away instead of being captured. But neutron capture is inevitable, and the result is invariably a secondary nuclear reaction which propagates induced gamma and other radiation throughout the environment.

The newly created radioactive elements may live anywhere from a few seconds to several thousand years and present widely different degrees of hazard. On striking an atom of nitrogen in the air, for example, a neutron may convert it into radioactive carbon-14 which decays over a half-life of 5600 years. Equally, on striking a copper-63 atom (such as would be found in a tank's electronics system), a high-energy neutron may convert it into radioactive copper-62 which has a half-life of 10 minutes and decays into nickel-62.

It is important in this regard to distinguish between induced radiation and fallout. Fallout consists of radioactive fission products as well as irradiated dirt, dust and débris thrown into the air by the blast and caught by wind patterns and the jet stream. Consequently, it will drift back to earth hours, days or even weeks afterwards, depending upon variables of wind speed and direction. Neutron weapons do create fallout, but at minimal levels.

Radiation kills largely through disrupting the delicate chemistry of living tissue through ionization — by altering the normal pattern of positive, negative and neutral charges in the atoms and molecules of the body. If the changes are severe enough, the organism will die before its biological system has time to repair the damage done. There is no set lethal dosage. Several micro-organisms and insects can survive radiation levels several hundred times that necessary to kill a man. Equally, the rate of exposure is important: one amount may be non-lethal for a sudden exposure but deadly for a constant exposure over a period of time, and vice-versa. Research has shown that the radiation energy needed to produce genetic mutations in human cells may, in some cases, be several hundred or even a thousand times less than for equivalent changes due to chemical causes.

The chief measure of radiation effects in man are the 'radiation absorbed dose' or rad and the 'roentgen equivalent man' or rem. These terms are defined in note 12. The statistical measure of lethality is usually the LD_{50} or the amount of toxic agent which is expected to kill 50 per cent of an exposed population. For man, the LD_{50} for radiation is about 450 rems over one week and 600 rems for an exposure over one month. The LD_5 (5 per cent expected deaths) is around 250 rems for exposure over a week and 350 rems over a month. The thresholds for radiation sickness are 150 and 200 rems respectively but long-term damage may be done at much lower levels. Five rems per year is the maximum level of full-body exposure at present permitted for industrial workers in Britain.

The timing and type of radiation death depend upon the levels received. An exposure of about 4000 rads would lead to death in two or three days. For dosages well above 4000 rads, death may be virtually instantaneous. At these levels of exposure, death results mainly from cerebrovascular damage affecting the nervous system and the heart and blood vessels. Vomiting and diarrhoea will begin from a few minutes to an hour after exposure and will be followed by a drop in blood pressure, delirium, coma and death. For exposures of roughly 2000 up to 4000 rads, death is mainly from gastro-intestinal damage and occurs around 10 days afterwards. The pattern of symptoms is the same.

Between about 150 and 2000 rads, death is not automatic but when it does occur it is caused mainly by damage to stem cells in bone marrow and lymphatic tissue. Death may take up to a month and the early symptoms will be followed by a dormant period where the patient appears to recover. Then damage to bone marrow and loss of white cells in the blood bring progressive weakness and vulnerability to infection — which may be the actual cause of death. The body areas affected are not mutually exclusive. A full exposure of several thousand rads will cause all the types of damage mentioned, but cerebro-vascular death will occur before damage to the other areas takes effect as bone marrow and white blood cells are most sensitive to radiation.

A rough rule for calculating the effects of induced radiation and fallout is that the total radioactivity will drop to a tenth of its former level for every seven-fold increase in time. This is really only relevant to civilians surviving a nuclear attack or troops entering areas contaminated by ordinary nuclear weapons. The vast majority of victims of neutron weapons would receive virtually all their irradiation within a few seconds through neutron bombardment and secondary, gamma-yielding reactions in the environment and the materials of their vehicles. The device would be air-burst for maximum area coverage by radiation and to minimize contaminating fallout. Those at the centre of the explosion would be killed by blast and heat while those 275 metres (300 yards) away would receive lethal bombardments of radiation equal to several thousand rads. Beyond that, however, the levels of radiation decline dramatically until, at the maximum extent of the weapon's radius (roughly 1370 metres or 1500 yards), dosages received would be reduced to the minimum threshold for radiation sickness or even below. After about 1370 metres there is the possibil-

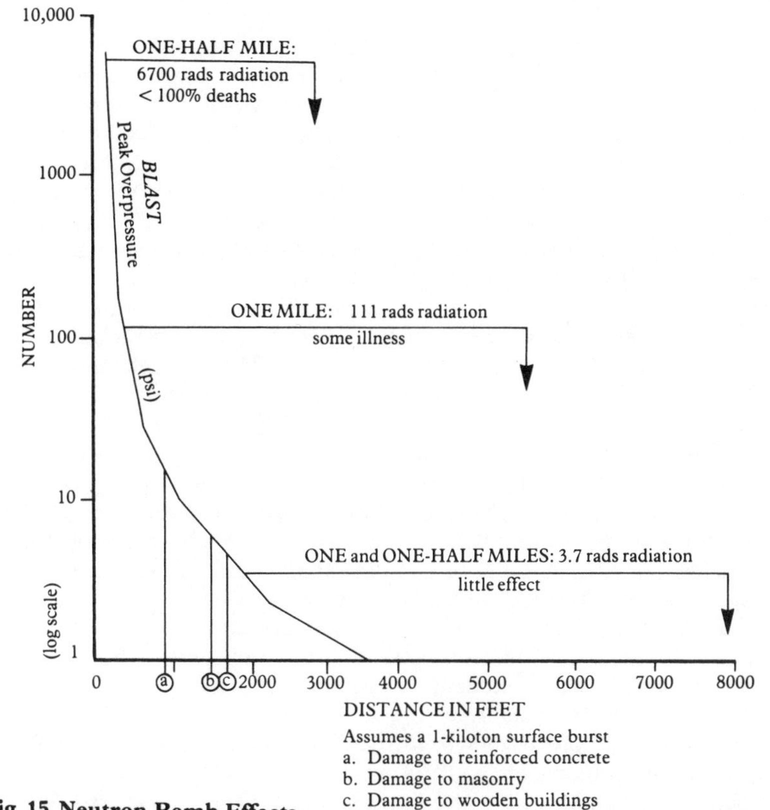

Fig. 15 Neutron Bomb Effects

ity that victims could suffer long-term damage but this does not enter into calculations of the weapon's immediate military effect. Victims would be likely to show higher than average incidence of leukaemia in later years, for example, or suffer genetic damage.

Because a very large proportion of neutron weapon victims would take between a few days and a month to die, critics have argued that use of the neutron bomb would invite a war fought by Russians driven to fanatical military performance by the certainty of death. Some survivors would even appear to recover and could be returned to temporary duty without reduced efficiency. But all this raises the intangibles of motivation under mental stress which cannot be predicted in advance. The point is important, however; neutron weapons have a measure of uncertainty about them. While the weapon's effects in terms of radiation exposure and consequent death rates are predictable, the time spans involved between exposure and death create doubts as to its immediate effects and therefore its exact military value.

Originally, neutron weapon designs were intended for deployment in two tactical delivery systems: a warhead (the W70–3) for the short-range Lance missile and a shell (the W70–9) for the 203mm (8-inch) howitzer. At a later stage, work began on a neutron shell for the 155mm howitzer which, at the time, formed nearly 75 per cent of America's artillery in Europe. Shortly before he postponed neutron weapon production, President Carter was informed that the components could not be miniaturized for the 155mm shell. His decision may therefore have reflected purely practical considerations rather than public opinion. The Army proceeded to develop a new, extended-range 155mm nuclear shell for European deployment which passed without comment; ironically, it was 'dirtier' than the weapon it replaced.

America's initial production order calls for 380 neutron warheads for Lance missiles (and, presumably, Lance's T-22 replacement tactical rocket) and 800 203mm (8-inch) artillery shells. The absence of a 155mm shell suggests that miniaturization problems have yet to be solved and this reduces the ERRB's flexibility.

The Soviets have been hinting at their own neutron weapons ever since Reagan's announcement and there is no technical reason why they could not begin production at any time. Soviet physicists have been conducting experiments with low-yield fusion devices since the early 1950s and some Russian writings on tank warfare indicate a definite appreciation of the possibilities of enhanced-radiation weapons. But, for the moment at least, their usefulness to the Russians lies chiefly in exploiting their propaganda value. Whatever military utility they have, neutron warheads are only obviously advantageous to a power outnumbered in conventional armour. At present, the Warsaw Pact has a tank advantage of three or four to one and this makes it prudent for Moscow to try and prevent neutron basing in Europe rather than legitimize Reagan's decision by deploying their own.

Whatever their mechanism and however limited their effects, ERRBs are nuclear weapons and their use would involve crossing the 'firebreak' dividing nuclear and conventional war. Restricting a nuclear war to the battlefield is qualitatively different to limiting a war to the use of conventional weapons. Either would be immensely difficult but the use of any nuclear weapon would be inherently escalatory because it would convert the second problem into the first one. By treating neutron weapons as a special type of anti-armour technique, the Pentagon has risked blurring the distinction between nuclear and conventional warfare. In general, spokesmen talk as if neutron weapons and ordinary nuclear weapons were distinguishable in principle because they have differences in effect. Capability and objective are stressed more than the physical nature of the weapon. This itself runs contrary to one of the central tenets of flexible deterrence — linking the level of response to the degree of provocation — which was one of the initial justifications of neutron weapons.

Radiological Weapons

Because it is designed to kill chiefly through radiation, the neutron bomb is strictly speaking a radiological weapon but a radiological warfare normally refers to the military exploitation of radioisotopes. These have been written and spoken about for years but few weapons have entered service. Neither of the superpowers has admitted deploying any and a treaty banning them has been under discussion for some time. For a time the US Army's Dugway Proving Ground in Utah had an area set aside for tests with radiological agents and the CIA included them in its researches into assassination agents.

When the Israelis went to war in 1967, they reportedly sent President Nasser a message threatening to blow up the Aswan Dam and to turn the Nile radioactive if he used his few German-designed rockets on Israel's cities. Before the Normandy landings, the Allies feared that the Germans might use lethal radioisotopes to contaminate the beaches and poison the ground for the duration of the agent's half-life. Strontium-90, for example, has a half-life of 28 years, is a potent beta emitter and is prone to deposit itself in bone marrow where it is highly lethal. Cobalt-60, tested at Dugway, is an extremely energetic radiator of gamma and beta radiation and has a half-life of just over five years. Unlike chemical or biological agents, radiological agents cannot be neutralized. Countermeasures such as soil clearance and other decontamination procedures — as well as the natural effects of wind and rain — would almost certainly clear the area long before the lifetime of the radioisotope came to an end, but how soon and how effectively would be uncertain. Long-term ecological damage would be virtually certain.

'Super-dirty' nuclear bombs coated with cobalt or other elements to produce deadly radioisotopes in the explosion have been discussed for years but it is not known whether such weapons have yet been stockpiled. However, the radiological option does not require having nuclear weapons at one's disposal, and any country with access to a nuclear reactor might develop a capability for radiological warfare. Nuclear reactors create waste products as the fuel is consumed through controlled fission. These waste products include any number of highly lethal radioactive materials; strontium-90, strontium-89 and cesium-137 are just three. How this waste should be disposed of is controversial in its own right, but it certainly provides a crude yet highly effective source of material for radiological weaponry. The materials could be inserted into specially shielded bombs, aerosol generators or rocket warheads and dispersed over enemy territory. Radiological devices based upon nuclear waste would probably be used strictly as revenge weapons to poison an environment rather than as battlefield instruments. Long-lived lethal radioisotopes could be used militarily in area-denial tactics — sealing off an area by contaminating it — but persistent toxic chemicals would be a more practical choice because their effects would dissipate within a few days or weeks at the most.

Efficient military weapons based upon radioisotopes have a unique set of problems to overcome. They require relatively short-lived materials which would have the desired lethal effects but not contaminate the area for unduly long. This imposes severe restrictions upon production. If the maximum length of contamination desired was a week, for example, a lethal radioisotope with a half-life of one or two days would be selected. But once the agent was produced, it would have a short shelf-life compared to toxic chemicals which may be stored for months or even years before use. Yet selecting radiological agents with relatively long half-lives for efficient storage would mean longer periods of contamination. In effect, the longer the half-life the more desirable the agent is from the viewpoint of production, storage and transport but the less its battlefield suitability. The shorter the half-life, on the other hand, the greater the battlefield efficiency but the more absurd the demands upon production.

One solution might be to produce the radioisotope on-site when required (as is done in research laboratories and hospitals) but this does not seem at all practical at present. The only practical solution would be to produce the radioisotope literally at the point of use in a nuclear explosion. Hydrogen bombs using uranium-238 for the wall of the fusion cylinder add a second blast-enhancing, fall-out maximizing fission reaction to the first fission-fusion chain. Lethal but short-lived radioisotopes might be produced in the same manner. Metals would be selected for inclusion in the bomb's materials on the basis of their ability to produce significant proportions of the desired radioisotope when struck by neutrons escaping from the nuclear explosion. Neutron bombardment, for example, may convert ordinary aluminium-27 into radioactive aluminium-28. Aluminium-28 is an extremely potent beta and gamma emitter but has a half-life of only 2.27 minutes. In itself, this is probably too short to have any military value, but in a proportion of the cases the aluminium-28 will decay into radioactive sodium-24 with a half-life of 15 hours. Sodium-24 is another extremely potent beta and gamma emitter and has a half-life which makes it militarily practical. It may also be produced from magnesium-24.

A weapon designed to maximize the production of sodium-24 might be an ideal radiological weapon. It could be used to create a cloud of radioactive dust that winds could disperse downwind to kill people without extensively damaging buildings and industry. Although other types of fallout would confuse the issue somewhat, the attacker would know when the level of radioactivity had dropped to a safe point. There are any number of radioisotopes which might be used in this way. Two other extremely powerful gamma and beta emitters are choline-38 with a half-life of just over 30 minutes and silicon-31 with a half-life of just under three hours.

The possibility of thermonuclear weapons which do not require fission triggers would make this sort of radiological technique more viable. Small-yield hydrogen weapons with their copious output of neutrons could be used to produce maximum amounts of lethal but short-lived radioisotopes without any of the long-range contaminants of the fission trigger. They would create some induced radiation but this could be minimized through air-bursts which had little blast effect on the ground but still released the radioisotope to drift over a city. Soldiers equipped with sophisticated protective clothing against chemical, biological and radiological agents would not suffer much additional hazard from short-lived radioisotopes in the environment. In the context of modern warfare, it is even doubtful if any major power would be especially concerned to bother stockpiling nuclear weapons which depopulated cities without destroying them but, if it did, 'clean' radiological weapons would be ideal.

8 Biological and Chemical Options

Practical Considerations

In 1347 the Mongols besieged the virtually impregnable Genoese trading fortress of Caffa on the Crimean peninsula. For some years, the plague had been spreading westward from Central Asia and it soon began to ravage the attackers' camp. As much in desperation as anything else, the Mongols began to catapult the corpses of plague victims over the walls and within a short time the defenders were suffering from a major epidemic. The Genoese took to their galleys and fled into the Mediterranean carrying the plague with them. By the spring of 1348, the Black Death had taken hold in southern Europe's major ports and was spreading inland rapidly. In the course of the next year, 20 to 30 per cent of Europe's population died from the plague; some writers have put the figure at closer to 50 per cent.

Writing of the plague, Alfonso of Cordoba referred to men wishing 'to do evil' who might break glass flasks and allow the winds to carry the escaping 'vapours' toward those they wished to kill. He thereby showed a reasonable grasp of what we have come to think of as chemical or biological warfare — an ancient practice which has only come to the fore with the pressures of military-orientated science in the twentieth century. When Richard the Lionheart made his second attempt to take Jerusalem in 1192 he found that Saladin had poisoned the nearby wells and, faced with an over-extended supply line, he withdrew. Louis XIV of France is said to have given a lifetime pension to an Italian chemist on the condition that he never revealed the secret of a biological weapon he had formulated. In colonial America, the British are reported to have exploited a smallpox outbreak at Fort Pitt, Pennsylvania, and distributed infected blankets to hostile Indian tribes.

Although poisons, tainted arrows and the like have been used for many centuries, it was not until 22 April 1915 that the full military capabilities of chemical warfare were exploited. On that day, at 5 pm, the Germans released nearly 200 tons of chlorine gas upwind from the French lines at Ypres. Some 5000 soldiers died and a gigantic gap was opened in the British trenches which the Germans — not knowing what to expect — failed to take advantage of. The Germans are also alleged to have infected Allied horses with glanders, and to have tried spreading plague on the Eastern and Western Fronts. By the time the conflict ended, chemical warfare had been used so widely that revulsion against it grew and eventually in 1925 a Geneva Protocol banned its use. All major powers are signatories but the United States and the Soviet Union interpret it as prohibiting a first use only and reserve the right to retaliate.

Despite numerous opportunities and the Germans' secret discovery of the highly lethal nerve gases sarin and tabun, neither side used toxic chemical warfare in World War II. On one occasion in July 1944, Churchill asked his Chiefs of Staff to report on the viability of using mustard gas and phosgene on Germany's population in retaliation for V-1 attacks and because of (unfounded) fears that the V-2s might devastate London. He was mainly interested in mustard gas, partly because the vast majority of casualties would probably recover eventually. He remarked: 'We could drench the cities of the Ruhr, and many other cities in Germany in such a way that most of the population would be requiring constant medical attention. . . . I do not see why we should always have the disadvantages of being the gentleman while they have all the advantages of being the cad.' Churchill's idea received little or no support and he quickly dropped it after the Combined Chiefs had rejected chemical warfare as having little military effect.

Since 1918 the only proven incidents of chemical warfare have been where modern armies used it on poorly equipped and ill-disciplined ones: the Italians in Ethiopia, the Japanese in China and the Egyptians in the Yemen. More recently, unsubstantiated allegations have added the Vietnamese in Cambodia and the Russians in Afghanistan. During the Vietnam War, America used the now highly controversial Agent Orange to defoliate the jungle. Agent Orange is a 50/50 mixture of the herbicides 2,4-D and its chemical relative 2,4,5-T which contains as a by-product the extremely lethal compound dioxin. There is now clear evidence that dioxin may cause long-term genetic damage even in sub-lethal doses and the American government is facing claims from former soldiers who were exposed to Agent Orange. There is also evidence of birth deformities in Vietnam associated with parental exposure to herbicides. The Americans also used cacodylic acid on rice crops against villages deemed to support the Viet Cong.

In September 1981 US Secretary of State Alexander Haig officially accused the Vietnamese of using Soviet-developed chemical agents in Cambodia, Laos and on the Thai border. Leaf samples were said to contain significant traces (up to 20 times any natural concentration) of agents extracted from poisonous fungi not generally common in the area. These chemicals were claimed to be the active ingredients of the 'yellow-rain' weapons reported in South East Asia since 1976. In the American view, yellow-rain is clearly a Soviet product since the prerequisite fungi are prevalent in the USSR. The accusations were not limited to fungal-based chemical weapons but, at that time, these were the only agents where specific supporting evidence was claimed.

During World War II biological warfare was practised by the Japanese in China and by partisans in Russia, Poland and Yugoslavia. Since the war there have been unsubstantiated allegations of American uses of biological weapons in Korea, Vietnam and — most recently — in clandestine attacks on Cuba. The Japanese wartime programme lasted from 1936 to 1945 and involved considerable human experimentation at a special facility constructed at Harbin in Manchuria. After the war the Americans obtained the data from these experiments in exchange for granting immunity from war crimes trials to several Japanese 'researchers'. A report of 1947 noted the value of the Japanese data and comments: 'Such information could not be obtained in our own laboratories because of the scruples attached to human experimentation.' There is now definite evidence that the Japanese also carried out several attacks with plague on Chinese civilians in an effort to start lethal epidemics.

In 1969 President Nixon unilaterally ordered the destruction of all America's stocks of biological agents and announced that the United States would never use biological weapons, even in retaliation. At the same time, he suspended the procurement of new chemical weapons, although old stocks were retained. In 1972 the major powers signed a Geneva convention prohibiting biological weapons, the stockpiling of agents, and development not justified by peaceful needs. However, any party may withdraw from the convention with six months' notice. Efforts to secure a similar treaty prohibiting chemical warfare have been going on since 1968 but have constantly foundered on the problems of verification and inspection. Moreover, the biological warfare convention carries no provision for either and is effectively a gentlemen's agreement.

Both biological warfare and chemical warfare include lethal anti-personnel agents, incapacitating substances designed to render the victim incapable rather than kill him, and weapons that can destroy crops and vegetation. Biological warfare also includes lethal 'land-wasting' diseases like rinderpest which are specific to animals, as well as agents such as anthrax which may affect both animals and humans.

The difference between 'incapacitating' and 'lethal' in both chemical warfare and biological warfare is not absolute. Incapacitating biological agents are not expected to be fatal in more than 2 to 5 per cent of the population exposed to them. Chemical incapacitants, which cover a wide range of physiological and psychological-orientated compounds as well as harassing anti-riot gases such as CN and CS, are not expected to kill in any but the most unusual circumstances. Nevertheless, if an incapacitating or riot gas is used to put a person into a position where he can be killed, the agent is, in a sense, lethal. Toxic chemical agents include lethal, highly poisonous nerve gases and casualty weapons such as mustard which usually incapacitate only, although they may also cause appalling physical damage.

Toxin Agents

The clear scientific distinction between 'chemical' and 'biological' does not always apply in the way agents are classified and discussed. During World War II, for example, synthetic herbicidal chemicals were categorized as biological weapons along with crop-destroying fungal diseases. The 1972 biological warfare convention includes a class of chemicals which are extracted and purified from biological entities but are themselves not living and, in some cases, may be duplicated synthetically in the laboratory. These include the deadliest organic materials known and are known as 'biotoxins' or often just 'toxins'. They fall into three main subdivisions: phytotoxins which are vegetable-based, zootoxins and microbial toxins. Zootoxins are found in animals; snake venoms are the classic example but certain species of frogs, toads and fish are sources for some of the world's most deadly poisons. Microbial toxins are found in fungi and bacteria. With a few important exceptions, toxins based on protein are the most lethal but they also tend to have higher molecular weights, which proved a drawback in trying to turn them into battlefield weapons.

Although they underwent military research and development, toxins became the concern of intelligence agencies and military units involved in the wide area of special operations. In practice, this has meant the use of exotic toxins for assassinations or in suicide gadgetry for captured operatives. Although the 1972 convention forbids the stockpiling of toxins for military or covert purposes, the amounts necessary for such things as assassination are so small that it is impossible ever to be certain that no signatory is cheating. In fact, in the early 1970s unauthorized toxin stocks were found in the CIA's possession, and there can never be any firm guarantee that small stocks are not still held by any of the world's clandestine agencies. The allegations that the USSR has used fungal-derived chemical weapons are, however, the first firm claim that toxins have ever been deployed on any large scale. Up to then, all the available evidence has suggested that, despite their extreme toxicity, they have never proved to be cost-effective for weapons and that their disadvantages outweigh their potential military benefits. Should the Russians be guilty, however, the Soviet Union would be in clear violation of both the 1925 Geneva Protocol and the 1972 biological weapons convention. Then the prospect of a renewed stockpiling of American toxin or even biological weapons is likely in face of public opinion and a discredited biological treaty, especially if research succeeds in synthesizing militarily viable products chemically identical or related to the more deadly toxins. Even more probable is further support for American chemical rearmament with new varieties of lethal nerve gas.

Nothing better illustrates the practical use of toxins than the assassination of Bulgarian defector and BBC broadcaster Georgi Markov in London in September 1978. Following his mysterious death, CID investigators found he had been attacked by a pinhead-sized pellet composed of 90 per cent platinum and 10 per cent iridium. The use of such rare metals pointed to a toxin as the cause of death because many naturally-based agents react adversely to more common materials. The 1.52 millimetre pellet contained two cavities of 0.35 millimetres, each of which were designed to contain one or more lethal agents. It was designed to be fired from an air gun which, in Markov's case, was probably disguised as an umbrella. After penetrating the skin, the pellet was activated by the victim's body heat and muscular movements which wore away a protective coating and released the agent into the bloodstream. A similar attack had been made on Vladimir Kostov, another Bulgarian defector, a month previously in Paris. On that occasion, however, the pellet seems not to have penetrated sufficiently to enable it to become effective. Both attacks are generally believed to have been carried out by the Bulgarian secret police in retaliation for outspoken criticisms of the regime and revelations of scandal within the ruling clique.

The agent used in the Markov killing was discovered to be a castor bean-based poison known as ricin. Ricin has been widely known and studied for years. It is one of the most deadly known poisons in the world and is some 25,000 times as toxic as strychnine; 1.5 micrograms can prove fatal to an average adult. It was only recently, however, that researchers discovered the mechanism by which ricin kills, which is that it inhibits the cells that synthesize protein. Even now, the mechanism is not completely understood but work undertaken at the Institute for Cancer Research in Oslo has found most of the answers.

Ricin apparently consists of glycoprotein bands which divide into two peptide chains (A and B) upon attacking a cell. The A chain (the 'effector'), and possibly the B chain (the 'haptomer') as well, enters the cell and carries the toxic constituent into the cytoplasm. The main function of the B chain seems to be to bind the ricin to the cell's surface and allow the A chain to enter. As in Markov's case, the symptoms of ricin poisoning are a rapid rise in the white blood-cell count followed by death due to toxaemia. There is no known effective antidote and ricin is extremely difficult to trace in a post-mortem. It has been widely studied throughout the world and was investigated for a variety of reasons including the detoxification of the plant from which it is derived in order to produce animal feed; it has also been studied as a possible cancer treatment because it appears to attack malignant cells more readily than normal ones.

Ricin has been considered a candidate warfare agent since World War I and has the code-name WA in the American arsenal. It has never been deployed in large quantities despite intensive secrecy surrounding its possible applications in what the US Army called 'toxicological warfare'. Nevertheless, there are any number of published methods for preparing ricin in small crystalline quantities suitable for laboratory work or clandestine assassinations. The US Chemical Corps' brief was to produce it in large quantities suitable for insertion into munitions and without loss of toxicity or stability. In a watery solution, ricin becomes unstable at temperatures above 60 to 75 degrees Centigrade and, in a solid form, after about 100 degrees. It is also sensitive to both acids and alkalis and may have its toxicity reduced by the heat generated in mechanically grinding it down into the particle size suitable for chemical warfare.

One approach was to grind and press the beans to remove most of the castor oil and then soak the resulting cake in a solvent to take out the small percentage of remaining oil. The cake was soaked again in a solution of acidified water to remove most of the toxic material and filtered. It then went through a long sequence of further soaking in a neutralized salt water solution, recovery, washing, filtration and the eventual separation of concentrated ricin. Next the mixture was dried and passed through a fine screen and air-ground to reduce it to a powder composed of individual particles with an average size between 2.5 and 3.5 microns. Chemical and biological warfare research had previously shown that this was the ideal particle size for attacking the lungs, where biological and toxin agents are most lethal, as smaller particles are likely to be exhaled without causing damage while larger ones are unlikely to reach the lungs.

The toxin which has received most military attention since the 1940s is the A strain of the bacterium *Clostridium botulinum*, designated by the code-name X. Botulism, which is most familiar as severe food poisoning, has several toxic strains. All have been extensively researched in Britain and the United States and the A strain, the most toxic, is preferred. Canada was also deeply involved in work on botulin and produced cruder but more stable varieties suitable for insertion into actual weapons. The bacterium itself has never been suitable for warfare because it is rapidly killed in the presence of oxygen, but it has nevertheless enjoyed a sensational press.

Botulin toxin A, which kills by a mechanism similar to ricin, is the most lethal organic substance in the world. In its purest form it is more than 660 times as deadly as ricin when injected in laboratory animals and well over 16 million times as toxic as strychnine on a unit-for-unit basis; in theory, around 28 grams (1 ounce) would be more than enough to kill every man woman and child on earth. (Comparative toxicities are detailed in notes 13 and 14.) It is exceeded in lethality only by radioactive alpha-emitters such as radium, polonium and plutonium. Nevertheless, it did not prove to be a suitable weapon material as it is relatively unstable in the presence of heat and a number of chemical substances. It was also comparatively expensive to produce; one estimate put the cost at around $200 per pound compared to a few dollars per pound for the less toxic nerve gases which are also more stable.

The third major toxin agent developed at roughly the same time was saxitoxin, coded TZ. Saxitoxin is a non-protein neurotoxin originally derived from California mussels and Alaska butter clams. The actual source is the plankton *Gonyaulax catenella* which turns the shellfish poisonous when they feed upon it. While saxitoxin may be extracted directly from the plankton, clams and mussels have generally proved easier to work with. It may be prepared by grinding clam siphons or mussel livers and mixing them with celite-545, ethanol and hydrochloric acid. The crude extract is then filtered and purified. Mussels are the most productive source but their period of maximum toxicity is limited to a few days while clams may remain toxic for up to a year, although their poison content is less. Saxitoxin is much less toxic than ricin or botulin and is comparable to nerve gas in its lethality, but it is a much more fast-acting and stable poison than either of the other two.

Much of the work in researching and later synthesizing saxitoxin was done by Dr Edward J. Schantz, who also did considerable work on botulin toxins and the incapacitating toxin agent PG.

PG (*Staphylococcal enterotoxin B*) is a white fluffy powder when freeze-dried and a highly soluble, simple protein containing amino acids only. It is derived from the bacterium *Staphylococcus aureus* which produces a common and generally non-fatal form of food poisoning in man. Its symptoms include violent vomiting, diarrhoea, fatigue and cramps. When it is fatal, respiratory damage is often involved and infection may also lead to pneumonia in weakened lungs. The US Army found that an effective dosage of PG was 0.1 micrograms per kilogram of body weight for tests on monkeys injected intravenously.

In measuring toxicity the effective dosage — ED_{50} — like the lethal dosage — LD_{50} — is the amount expected to induce the desired effect in half the animals tested. A synthetic or toxin agent's suitability as an incapacitant is indicated by its safety ratio, which is found by dividing the ED_{50} by the LD_{50}. The larger the number, the safer the agent. With very low numbers, there is no practical distinction between incapacity and lethality because the agent will have a high chance of killing. Serious candidate incapacitants have safety factors in the range of 100 or above. BZ, for example, a synthetic chemical which produces hallucinations, has an ED_{50} of between 5.7 and 6.7 micrograms per kilogram of body weight and a LD_{50} between 0.5 and 3.0 milligrams per kilogram; the safety factor ranges between 87.7 and 447.7.

If the American accusations of Soviet use of fungal-based 'yellow-rain' chemical weapons are accurate, four specific toxins are involved: T2, DAS, nivalenol and deoxynivalenol. All are among the toxins produced by varieties of *Fusarium* fungi such as *Fusarium tricinctum* of the trichothecene group. T2 toxin has also been isolated from *Trichoderma lignorum* which occurs in mouldy corn. Symptoms of trichothecene poisoning include vomiting, blistering and internal haemorrhaging, while death is associated with lesions and haemorrhaging of the intestines, liver and kidneys. *Fusarium* fungi occur widely and are known, for example, for polluting corn in the American state of Wisconsin. Both Russian and American research into *Fusarium* toxins is extensive and at least one US study was partly funded by the Department of the Army as well as the US Public Health Service.

However, the trichothecene toxins are relatively weak and this has caused some doubt as to their development into chemical weapons. Death from trichothecene poisoning is often prolonged and requires considerable dosages over an extended period of time. In one experiment, 100 micrograms of T2 were injected into a steer for 65 days before it died. Studies involving exposure of young trout to another trichothecene toxin indicated a LD_{50} of up to 6.1 milligrams per kilogram. Effective dosages, however, are lower and if the Soviets have gone to the trouble of developing trichothecene toxin weapons they are probably designed as incapacitants or rice-pollutants. According to the varied and often ambiguous reports, many of the supposed chemical attacks have involved aircraft-fired rockets emitting red or yellow smoke causing vomiting, diarrhoea and convulsions but few deaths. Other reports speak of 'hundreds' of deaths but it is not claimed that trichothecene poisons are the only agents being used. Some scientists have also objected to the idea that *Fusarium* are naturally absent from Southeast Asia and have pointed out that little if anything is known about the fungi's distribution in the region.

Some toxins have attracted at least a brief interest from military researchers at one time or another. Officially, today's military-oriented toxin and biological warfare research is not for making weapons but for identifying ones an enemy may adopt. However, toxin research may provide fruitful lines of enquiry for the synthesis of chemical warfare agents — either a synthetic version of the toxin itself or its chemical homologues and analogues, which sometimes prove to be more biologically active than the original. The natural sources of many exotic poisons are often too limited to provide agents for mass deployment — batrachotoxin, for example, is a highly lethal non-protein poison some four and a half times as deadly as saxitoxin. It is used by the Indians of the Choco rain forest in Colombia as a tainting agent for blow-gun darts. When the poison first began to attract serious interest by toxicologists, the only source was the skin of the South American arrow poison frog. During the first eight years of work, four expeditions were sent to what was described in 1971 as 'the impervious jungle' of western Colombia to obtain supplies. The practical difficulties were great, the frogs did not travel well and the toxin, once obtained, was found to be very difficult to work with. However, a batrachotoxin derivative has been synthesized with a toxicity of 1 microgram per kilogram compared with 2 micrograms for the natural product.

Another agent which has attracted considerable military interest is tetrodotoxin, and this has had the advantage of a reasonably accessible source. It is extracted from the viscera and sex organs of certain species of puffer fish. From a scientific point of view, one of the toxin's most intriguing features is its chemical identity with tarichatoxin, a poison found in a species of California newt. Tetrodotoxin (also known as fugu poison) is a public health problem in Japan, where the puffer is regarded as a delicacy, and much of the considerable work on it comes from Japanese researchers. Tetrodotoxin's mechanism of action is similar to that of local anaesthetics (it has legitimate medical applications) and death is due to respiratory failure.

Curare's suitability for military purposes has been investigated since 1941, when the United States and Britain began taking a serious interest in it after receiving reports of German work. Research soon found that a mixture of finely divided crude curare and a nasal irritant had an almost immediate paralysing effect when released as an aerosol or powder. At the time, the relative scarcity of the material was a serious drawback although its covert use in darts remained a serious interest for a long time afterwards. Curare causes paralysis of the skeletal muscles and death is due to asphyxia. As a paralytic, it now has medical applications and military and CIA researchers early saw the possibility of using it in trauma-relieving medications for self-administration in the field. Curare-based preparations were also considered as possible field treatments for nerve-gas poisoning, but the final choice was a spring-fired unit of atropine for self administration. Atropine is also toxic, but much less so than curare.

The extent of the searches undertaken for toxins can be appreciated from an American Defense Department summary of work done under a contract with the University of California. It involved sending an expedition to the rain forests of Peru and Ecuador and gathering samples of 2500 plants to be examined for toxic or incapacitating effects. An early CIA memo described efforts to obtain the gall bladder of a diseased Tanganyikan crocodile in order to research a poison supposedly obtainable from it. Help from an unnamed zoologist and a local witch-doctor was assured but local law prohibited the transportation of diseased crocodile viscera. The recommended solution was therefore to obtain a live diseased crocodile and ship it to the United States for dissection. Unfortunately, the declassified documents do not reveal whether this worked or whether the rumoured poison ever proved to be worth all the trouble.

9 Biological Warfare

Biological Weapons: Theory and Practice

Biological warfare (BW) involves the use of disease-producing micro-organisms — bacteria, viruses, fungi and rickettsiae — in support of military or paramilitary operations. There are two general methods of delivery: direct or via a vector. A vector is an animal or insect which transmits the disease from an infected individual to a healthy one, and although there are exceptions, the intervention of the vector is normally necessary for the disease's transmission. The yellow fever virus, for example, requires a mosquito vector as it is parasitic on both man and insect and has a sophisticated life-cycle involving the two. Direct transmission, on the other hand, simply involves creating a sufficient amount of agent in the air to cause infection through inhalation or ingestion. This requires the creation of a cloud of infectious material over the area to be attacked. The cloud must be reasonably uniform; in other words, there should be a fairly even distribution of toxic particles in the required concentration. If there is not, only a few individuals within the attacked area will receive sufficient micro-organisms to be infected while some will receive more than necessary.

Pulmonary infection is by far the most effective, and intensive research has been done to prepare fillings for biological delivery systems which would distribute a relatively uniform cloud of particles in the optimal size of 1 to 5 microns. In most cases, the lethal cloud must be distributed as quickly as possible because the majority of biological agents are highly sensitive to environment variables such as heat and ultraviolet light. When released into the air, the number of viable organisms will rapidly begin to decline until the proportion of infective ones is unlikely to cause infection. The half-life of the cloud may be only a few minutes and, while some hazard remains, the chances of causing infection decline quickly with the rate of aerobiological decay. Because of the ultraviolet radiation in sunlight, night time has always been considered the best time for biological attacks.

There are many biological agents which will infect through wounds or abrasions in the skin but this has not been considered a primary means of attack because it would only create a hazard for troops already wounded or injured. Particular projects developed biologically contaminated artillery and mortar shells as well as fragmentation bombs to cause skin infection additional to blast and shrapnel effects but these are of less general interest than munitions designed to create lethal clouds challenging the lungs. Tainting conventional rifle or machine-gun ammunition with toxins is another approach but these special-purpose munitions are not intended for large-scale use.

Skin penetration is the most lethal means of delivering toxin agents but it is not practical for attacks designed to infect large numbers of people over a wide area. The theoretical minimum lethal human dosage of botulin toxin A, for example, is 0.00003 micrograms per kilogram of body weight when injected compared to an LD_{50} of about 0.3 micrograms if inhaled and 0.4 micrograms if ingested. Thus botulin toxin A is some 140 times less effective when inhaled and around 190 times less effective in the digestive tract than when injected. The overall effectiveness of biological weapons is therefore determined by two important considerations: the efficiency of the agent in causing death or incapacitation and the efficiency of the actual weapon.

In theory, a single micro-organism should be able to produce an infection once it has taken hold in the body and begun to reproduce. In practice, however, this rarely happens. Plague apparently requires around 3000 bacteria for infection and estimates for anthrax range anywhere from 20,000 to 100,000 bacilli. Many of the incapacitating agents, on the other hand, require only a very few micro-organisms to be effective. The Anglo-American standard for virulence seems to have been between 25 and 50 per cent fatalities for lethal agents and between 20 and 30 per cent incapacitation, with a very small percentage of deaths, for non-lethal agents.

Work during World War II showed that the effectiveness of air-delivered biological weapons was increased by using numerous small bomblets or aerosol-type dispensers rather than a few large devices, which tended to concentrate the agent around the point of dispersal giving uneven distribution and wasting material by creating concentrations far in excess of the required effective dosages. Saturating an area with numbers of small weapons, however, created individual agent clouds which were more likely to even-out into a single cloud with a more uniform distribution. The first modern biological weapons were explosive bomblets filled with a slurried mixture containing a percentage of micro-organisms.

The first of these weapons was developed in America during World War II from a British prototype. Based upon the design for a incendiary cluster weapon, the unit weighed 227 kilograms (500 pounds) and carried 106 Type-F bomblets weighing 1.8 kilograms (4 pounds) each. Each bomb was filled with a slurried mixture of 340 grams (12 ounces) containing 96 per cent water and 4 per cent anthrax bacillus. After the war the Americans developed an improved 4-pound bomblet to be carried in an E-96 cluster-unit of 104 bomblets each. The bomblets contained a slurried mixture of 320 cubic centimetres of biological warfare material and were designed to be filled with anthrax or botulin toxin A as well as a number of other agents developed since 1945. Released at an altitude of 10,700 metres (35,000 feet), the E-96 could operate at 7600 metres (25,000 feet) and saturate an elliptical area of 920 by 240 metres (3000 by 800 feet) with bomblets.

Later on, improved munitions were developed which did not depend upon the use of an explosive charge. They included hydraulic atomization of a liquid-agent filling by using gas pressure to force it through a nozzle so as to create a fine spray. Another pressure system forced solid 'dry' particles of agent out of a

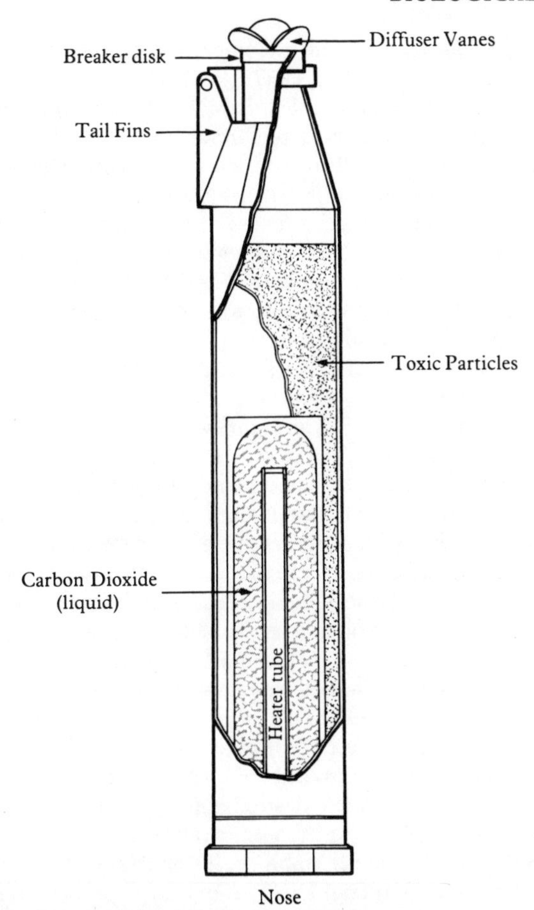

Fig. 16 Biological Warfare Bomblet (dry particles)

Impact with the ground causes a primer in the nose to ignite and detonate the heater tube containing potassium perchlorate, charcoal and oil. The burning within the tube generates a large amount of heat which vaporizes the liquid carbon dioxide contained in the cylinder surrounding it. Pressure within the cylinder breaks a rupture disk at the base and escapes into the compartment containing the toxic particles and up through the tail where it forces the agent out into the open air. The diffuser vanes divide it into three separate low-hanging clouds for maximum effect. The drop in pressure causes the heater tube to burn at a slower rate and avoids overheating the toxic particles.

container. A dry-agent bomblet using this last technique is illustrated above. Dry-agent payloads contained far higher proportions of micro-organisms than liquid fillings but some biological agents seem to have been more suited to slurried fillings. Dry-agent fillings also appear to have put far greater cost and safety constraints upon production.

Another method for distributing biological agents is from spray tanks carried by aircraft. Towards the end of 1957 North American Aviation received a government contract to research the possibility of launching large-scale attacks of biological weapons from low-flying aircraft. The idea was to use a pressure source to eject a liquid-agent filling through a fine nozzle. Droplets of 5 microns or less would be sprayed out and drift over a wide area. In the course of the following year, six trials were conducted at the Dugway Proving Ground in Utah. An F-100A with tanks fixed to its wings 'attacked' a triangular area 24 kilometres long by 39 kilometres wide (15 by 24 miles). Five tests involved the use of simulants but at least one employed live Q-fever and the results claimed that 99 per cent would have been infected if the attack had been carried out in a populated area. The Chemical Corps concluded that 15,140 litres (4000 US gallons) of agent dispersed by a single aircraft at night and with a wind-speed of 16 kilometres per hour (10 miles an hour) would cause half the population to become ill within an area of 130,000 square kilometres (50,000 square miles). Q-fever is highly infectious and requires only a few organisms for an effective dosage compared to other biological agents so these results, even if they were accurate, should not be taken as indicative of the performance of other agents.

Stability is another important consideration because agents have to retain their desired properties for acceptable periods of storage, during production and in use. Many selected biological agents have storage lives of a few months (much less than most weapons systems) although some are stable for longer. But storage usually requires fairly stringent standards of refrigeration. Stability in production has sometimes proved difficult because the process and the growth medium contribute to undesirable mutations. Equally, however, the production medium and process can be exploited for the intentional manipulation of mutations to breed desired qualities into a new strain of the organism, but this may also bring about unwanted secondary effects — a strain with higher virulence but lower stability, for example. It has been found that the means of dissemination could also render large proportions of the micro-organisms ineffective, since they are sensitive to such things as heat and blast.

The agents need to have a minimal immunological tolerance in the target population. Public health programmes and natural selection have reduced the number of diseases which would be suitable biological agents. One answer has been to breed strains for which there is little or no natural immunity and no available vaccine, such as anthrax and plague. Ideally, suitable vaccines would be available — but only to the user. However, this is a condition which is difficult to fulfil in a world where such scientific discoveries tend not to remain secret for long.

The time between infection and effect varies widely with various diseases, making a low incubation period an important consideration. For battlefield uses of biological agents the longer this delay the more impractical the weapon. This is the main reason why biological weapons have never been considered to be of much tactical use. But there are certain exceptions. During the 1962 Cuban missile crisis, a proposal to spray Cuba with an incapacitating biological agent — Q-fever, possibly in conjunction with tularaemia — was seriously proposed as a 'softening-up' exercise before an invasion. The attack would have been carried out with a sufficient time lapse before the actual invasion to allow the disease to take hold.

Most selected biological agents have been chosen for their low or non-existent potential for creating epidemics, as highly contagious diseases carry the problem of reciprocity and may eventually infect the user as well as the victim. However, epidemic-producing diseases such as plague have been intensively investigated, probably because of the fear that the Soviet Union might have gained an advantage in understanding the mechanisms by which human epidemics take hold.

There are any number of variables which will influence the extent and severity of an epidemic. They include the efficiency of the health services (curative and preventive), for example, and the degree of natural or induced immunity in the population at any given time. With vector-borne diseases, there is an ecological relationship between the human and carrier population, and social dislocation may lead to an increase in the number of vectors and a consequent rapid rise in incidence of the disease. Epidemic typhus and Venezuelan equine encephalomyelitis, for example, are not directly contagious but may be borne by fleas, lice and insects. These diseases will spread rapidly when the vector population increases because of some sudden imbalance in the natural ecostructure or breakdowns in sanitary conditions. But, contrary to popular belief, a predictable agent capable of producing epidemics among humans has so far eluded biological warfare researchers.

Biological Agents

If predictable epidemic-producing agents for human populations have proved elusive, the same is not true for animal or crop diseases. From the very beginning of the postwar programme researchers felt confident of their ability to cause catastrophic epidemics in domestic animal populations with such viral diseases as foot and mouth, rinderpest, wart hog disease, fowl pest and hog cholera. Each of these had a predicted mortality rate of between 20 and 90 per cent but the wide variation in estimates probably indicates a less complete understanding of animal epidemics than now-declassified documents indicate. As far as crop diseases were concerned, there was slightly less optimism over the ability to instigate epidemics since many of the agents required fairly precise climatic conditions for propagation. *Pyricularia oryzae* (rice blast), for example, attacks rice and other grassy plants and normally requires damp and humid conditions for germination and spread.

Crop-infecting fungi were considered the most important anti-crop agents after synthetic herbicides. Considerable work was done on the following fungi (their original British military code-names are given in brackets): *Sclerotium rolfsii* (C), *Phytophthera infestans* (LO), *Helminthosporium oryzae* (E) and *Pyricularia oryzae* (IE). *Sclerotium rolfsii* attacks soya beans, sugar beets, cotton and sweet potatoes, *Phytophthera infestans* is responsible for potato blight and caused the Irish potato famine of 1845 to 1875, and *Helminthosporium oryzae* causes brown spot disease in rice. After 1951 the Americans appear to have concentrated on *Pyricularia oryzae* and *Puccinia graminus tritici* (TX) which attacks wheat. Between 1951 and 1969, the US Army carried out at least 31 anti-crop tests, and rice and wheat blast fungi were stored at Fort Detrick and Edgwood Arsenal, Maryland, and the Rocky Mountain Arsenal in Denver, Colorado. Experiments showed that 3 grams of the rice blast fungi per hectare could infect between 50 and 90 per cent of the crops exposed.

Of all the human pathogens investigated, only a few ever became standardized and adopted into the American arsenal. The evidence, such as it is, suggests that this was also the experience in the Soviet Union. Among the agents adopted, anthrax (coded N) was one of the earliest and the most universally studied. Thought to have been the fifth plague of Egypt described in the book of Exodus, anthrax is a bacterial disease caused by *Bacillus anthrasis* which may propagate itself in a sporulation stage. In this form anthrax is extremely stable and is capable of persisting in soil for years, a persistence that is actually a disadvantage from a military point of view. Anthrax was also examined as an anti-animal agent but its anti-personnel potential was always paramount. The bacillus kills through skin abrasions, inhalation or ingestion and is not normally transmitted directly between individuals. Pulmonary anthrax was the principal technique exploited, although it was also studied as an ammunition tainter and a water contaminant. Pulmonary anthrax may be fatal in up to 100 per cent of untreated cases and has an incubation period of between one and five days, the lower figure representing extremely heavy doses of bacteria. At first the disease may appear as a mild cold but soon progresses to high fever, vomiting, increasing breathing difficulty, coma and death.

Undulant fever or *Brucella suis* (coded as US) is a highly infectious and disabling disease with no potential for causing epidemics. It is normally fatal in 3 or at most 5 per cent of cases and was a standardized agent for biological warfare in spite of a high degree of bacterial instability which had to be improved by the addition of protein products. The incubation period is between 10 and 75 days. Undulant fever causes a systematic debilitating infection which may last for a few weeks to a few months. Infection is through the respiratory or gastro-intestinal organs or through skin abrasions but, for military purposes, the first is by far the most important. It was the second major biological agent to receive attention after anthrax and was attractive because of its high infectivity and its tactical potential as an incapacitant. It also may infect animals but this does not seem to have been a line much pursued.

Rabbit fever or tularaemia (UL) is another highly infectious disabling disease with little or no risk of producing an epidemic. Its incubation period is one to fourteen days with three days being the most usual. The bacteria (*Francisella tularensis*) may be naturally transmitted by a number of flies or body ticks carried by rabbits and foxes, as well as by infected meat, water or animal hides, but was chiefly designed for aerosol dissemination. As a standardized agent, UL is unusual in that there was a wide divergence in the lethality of the strains available. Serious work on it began around 1958 when research reports referred to an incapacitating strain that affected the glands and caused fever prostration but would kill only about 5 per cent of those infected. By the time the agent was adopted by the military some 10 years later, the emphasis had switched to a typhoidal strain capable of proving fatal in 30 per cent of untreated cases; if, as is likely, pneumonia sets in the fatality rate rises to 40 per cent. Antibiotics were known to be highly effective in holding back the disease's progress but largely ineffective in modifying its disabling effects. This was considered crucial since convalescence could take up to half a year, turning soldiers into long-term hospital cases and putting severe strain upon health facilities as well as withdrawing men from military service. Compared with others, the UL bacteria is quite unstable (it can be killed by temperatures of 45°C) but the range of strains available and their varied lethality were enough to compensate for low stability. The variant chosen as a biological weapon was resistant to treatment by streptomycin.

Another incapacitating disease, Q-fever (North Queensland Fever), was coded OU in the American inventory. It is caused by the rickettsiae *Coxiella burnetii*. Rickettsiae are midway in size between viruses and bacteria but, like viruses, may only reproduce themselves within living cells. As a biological warfare agent, Q-fever is comparatively stable and extremely infectious with one micro-organism being capable of producing incapacitating symptoms if not a full onset of the disease. Its chief disadvantage is its comparatively long incubation period which may take up to three weeks. Thus it has been intensely studied in combination with other faster-acting agents. Q-fever is seldom fatal (about 1 per cent of those infected) and resembles a severe influenza which may last for up to a month without treatment and leave the victim incapable of normal performance for another month or two. Infection is typified by chills, fever and severe muscle pain with the possibility of an added complication of pneumonia.

Psittacosis or parrot fever (SI) entered America's military inventory for a short time but was no longer being held by the time President Nixon ordered biological warfare stocks to be destroyed in 1969. It is a viral disease resembling typhoid fever which may develop into pneumonia and prove fatal in about 20 per cent of those infected. Its incubation period is between one and two weeks and, once infected, the victim may need a month or longer to recover fully. The disease affects both humans and birds but the primary military route considered was the aerosol rather than the vector. Parrot fever is potentially epidemic, unlike the other agents mentioned so far. It attracted the attention of the CIA which investigated it as a possible contaminant for paper, letters, money and stamps, and a CIA memo refers to usage of parrot fever but does not give details. The memo also refers to other (unnamed) groups who have made 'intensive studies' of 'germ treated papers and germ treated items'.

One vectored agent included in America's biological arsenal was yellow fever (OJ), which is transmitted by the female mosquito of the species *Aedes aegypti*. The mosquito vector had its own code designation OJAP. The mosquito feeds upon animal or human blood every three days or so. If infected, it transmits the disease during feeding and may infect any number of individuals during its life-cycle of several weeks. The incubation period averages three days but may take up to ten. Symptoms appear suddenly and cause high fever, vomiting, headache and incapacitation with death — if it occurs — coming in about a week. The fatality rate may be 30 or 40 per cent in untreated cases. If the victim recovers, he may be severely weakened for two weeks to two months.

The chief attractions of yellow fever as a biological weapon appear to have been two: first, a belief that the population of the Soviet Union (which has never suffered from an epidemic of the disease) would be highly vulnerable; and, secondly, the persistence of the agent in the attacked area by virtue of the insect vector's lifetime, which in practice means several weeks compared to as little as a few minutes in the case of most agents released into the environment without the benefit of a live carrier. Tests indicated that several square kilometres could be covered by a few mosquito-containing devices and army engineers designed a plant capable of producing 130,000,000 mosquitoes a month. Weapons systems included containers weighing less than 1.3 kilograms (3 pounds) for a 340-kilogram (750-pound) cluster device, spherical bombs of 114 millimetres (4½ inches) for aircraft and the 89 millimetre (3½ inch) spherical bombs for the biological weapon warhead of the Sergeant tactical missile. Mosquitoes were infected by soaking larvae that were three or four days old in blood serum taken from infected rhesus monkeys.

Venezuelan equine encephalomyelitis (NU) was another potential vector agent that could be transmitted by the mosquito, but when it was finally adopted it was chiefly as an aerosolled agent. NU is severely incapacitating in man but only fatal in something like 2 per cent of those infected; aerosolized NU, however, is thought to be lethal in 3 or 4 per cent. This viral disease resembles severe influenza in man, has an incubation period of approximately four days and cannot be directly transmitted between individuals. Once infected the victim suffers a sudden onset of fever, vomiting, chills and body pain. The symptoms persist for a week or two and are generally followed by complete recovery. NU is highly infectious and requires only a few micro-organisms for an effective dosage.

Although it does not seem ever to have been standardized, plague has been widely researched in the United States, the Soviet Union and Britain since World War II. Caused by the bacterium *Pasteurella pestis*, plague comes in three forms: bubonic, septicaemic and pneumonic. Although bubonic was the most prevalent form, each of the three seems to have been involved in the pandemic of 1348. Both bubonic and septicaemic (the rarest form) require flea vectors carrying the disease from infected rodents or, in the case of septicaemic, from an infected individual directly to a susceptible one. The pneumonic form is by far the most deadly. Untreated, it may easily produce a 100 per cent fatality rate. It attacks the respiratory system and infection is directly spread by air-borne bacteria. There have been several accidents with pneumonic plague in

BW laboratories and, in 1962, a worker at the Microbiological Research Establishment at Porton Down died of the disease. It is moderately infectious and has an incubation period of approximately three days; symptoms are severe fever and chills followed by delirium or coma and death within two to four days, but sometimes even sooner.

Aerosolled forms of the bacteria were intensively researched but the organism is highly unstable once released into the air and may decay at a rate of 10 to 75 per cent in a minute under certain conditions of weather and daylight. Strains of the greatest interest combine virulence with resistance to streptomycin. Although there is no certainty that an epidemic would actually take hold on any particular occasion, the chances are too great to allow the use of plague in anything but terror attacks or sabotage. Plague is also an effective contaminant of water and food stocks where it may persist for longer periods, but anybody using it in this way would always have to be prepared for a possible epidemic.

To a very large extent, biological warfare research work is identical to that done in any public health or medical laboratory; the difference is one of intent rather than focus. This is one of the main reasons why the signatories to the 1972 Biological Warfare Convention were willing to forgo verification and inspection clauses. Although some work — studies of aerosol dispersal and weapon design, for example — would be clearly offensive and contrary to the treaty, much could as easily be interpreted as peaceful work and legitimate enquiry into pathogenic diseases on the grounds of public health. At the very least, a large proportion would be ambiguous and not necessarily evidence of ill intent.

BW: Myth and Reality

The minute amounts of toxin, viral and bacteriological materials required to kill or incapacitate has produced an almost universal belief that they constitute the basis of some of the most efficient weapons science has developed. As yet this is little more than myth created by a combination of governmental secrecy, dramatization by the media and the colourful imagination of thriller writers. This is not to say that there is no danger of biological warfare or that it would not prove deadly; quite on the contrary, for despite the 1972 convention there is still a considerable threat. But the real threat is not as might at first appear.

In 1980 and 1981 several biological warfare stories reached the headlines. Most of them were concerned with germ warfare accidents in the Soviet Union and warnings of clandestine Russian BW development in contravention of the 1972 convention. In March 1980 the US State department referred to an incident that supposedly occurred in April 1979 at Military Village 19 on the southern outskirts of Sverdlovsk, where the accidental release of anthrax was said to have killed 1000 or more people. This story was basically a repeat of a February 1980 report in London's *Daily Telegraph* which also told of another major BW accident during a Soviet troop training exercise in eastern Europe the previous autumn.

However, these suppositions have been built on flimsy evidence. In fact, if all the recent reports of Russian germ accidents were true, the Soviet Union would somewhat improbably have suffered three major biological warfare disasters within a single year. Testifying before the US Congress, Dr Mark Popovsky, a Soviet émigré since 1977, said he learned of the Sverdlovsk incident in a letter through the underground in February 1980 (the same month, coincidentally, that the *Daily Telegraph* reported the story). What happened, he said, was not an isolated event but 'proof of fifty years of Soviet biological warfare research . . . The secret compound at Sverdlovsk is one of a series of incidents and field sites where specialists work on such communicable diseases as the plague, tetanus, anthrax and yellow fever.' Unfortunately for Popovsky's case, none of the diseases mentioned (with the exception of the pneumonic strain of plague) is 'communicable' in the medical sense of the term.

Tetanospasmin, the chemical neurotoxin extracted from the tetanus bacillus *Clostridium tetani*, is one of the world's most lethal poisons but the bacterium itself is totally unsuitable for biological warfare because, like the botulin bacterium, it is killed by oxygen. Tetanus also requires a wound for effective entry into the body, where it acts on the synthesis of acetylcholine and inhibits the synaptic transmission of nerve impluses. Any Soviet work on tetanus would therefore be involved with the extracted chemical toxin — which is no more 'communicable' than arsenic — rather than the bacterium. Yellow fever and the bubonic strain of plague require a vector for epidemic spread and the chances of anthrax being transmitted directly between people are so limited as to be negligible. This is one of the principal reasons why anthrax was adopted as a biological warfare agent in the first place. There are certainly some legitimate grounds for suspecting a germ warfare accident at Sverdlovsk but they have become lost in the mythology that has surrounded biological warfare since World War II.

All the available evidence suggests that Soviet researchers into biological warfare followed the same lines and came generally to the same conclusions as those in America and Britain. During the Sverdlovsk controversy, Rear Admiral Thomas Davis, Assistant Director of the US Arms Control and Disarmament Agency, testified before the Senate on the possibility of Soviet duplicity. He spoke of an American multi-departmental study which had concluded in the late 1960s that biological weapons were 'an impractical instrument of warfare'. He then said of America's biological warfare policy: 'The United States did not give up biological weapons due to reliance upon the convention. Rather *having given them up, we wanted treaty restraints upon others.* It is in our interests to try to hold the Soviets to the convention, and to encourage adherence by other nations' (author's italics). There is no reason to doubt that the Soviets share that opinion.

The credit for biological weapons and the myths that surround them goes largely to Britain as declassified documents clearly reveal. In 1936 the British Committee for Imperial Defence was charged with examining defensive measures to biological attack, and three years later the Cabinet ordered a more thorough programme which included research into offensive techniques.

Although documentation is sparse, this work seems to have been prompted largely by a fear that the Germans were investigating biological warfare and reports that the Japanese were using BW (including plague) in China. A 1941 anti-crop campaign by the RAF against Germany was considered but discarded because of insufficient aircraft and the extension of the war into the fertile areas of south-eastern Europe. Anthrax became the responsibility of a bacteriological unit created at Porton Down in 1940. By late 1941 enough progress had been made to conduct an open-air test on the small uninhabited island of Gruinard off the west coast of Scotland. The tendency of anthrax to persist in the soil is so great that Gruinard is still classified as uninhabitable by the Ministry of Defence.

The results of the Gruinard test and a repeat in 1942 seem to have so impressed the British that the United States was actively encouraged to begin its own work. Serious American BW efforts began in 1942 with a number of secret contracts between the War Research Service of the National Security Agency and several universities. At the same time a Special Projects Division was set up within the Chemical Corps, and bacteriological facilities and a pilot production plant were planned at Fort Detrick. By the end of the war America had added a biological testing station at Granite Peak, Utah, another testing station at Horn Island off Pascagoula, Mississippi, and a large-scale production plant at Vigo, Indiana. In July 1944 Churchill requested feasibility studies for chemical warfare and any 'other method of warfare we have hitherto refrained from using'. This suggested a flirtation with using anthrax, although he had already been informed that sufficient munitions would not be ready until mid-1945 at the earliest.

A British document of November 1945 on biological warfare's *future* possibilities quotes a damage estimate for the 500-pound anthrax cluster and illustrates its point in terms of the number of weapons it would take to kill half the population of six German cities. According to this document, 60 cluster units (6360 4-pound bomblets) per square mile plus an additional 25 per cent to make up for ineffective bomblets would kill half the population of each, as well as creating the risk of additional deaths through skin infections and contaminating the soil for several years.

There is no suggestion that such an attack was ever seriously considered, however. British and American reports from 1945 to 1950 show that no BW capability ever materialized. America's Vigo plant was only 'essentially complete' by the spring of 1945 and only 'proven satisfactory for production' with an anthrax simulant (*Bacillus globigii*) by VJ-Day in August 1945. Even then 'great technical difficulties and dangers' remained and had yet to be fully overcome. The maximum capacity at Vigo was either enough botulin toxin A to fill 250,000 4-pound bomblets a month or sufficient anthrax for 500,000 per month. Although Britain received some experimental anthrax bombs for testing, postwar documents make it clear that anthrax was never produced in sizable quantities at Vigo or anywhere else. Even if Britain and America had planned to use anthrax on Germany, and had production begun around August

1945, the full six-city attack would not have been possible before the spring of 1946 at the earliest. To produce the total anthrax filling necessary, Vigo would have had to operate at full capacity for more than eight months without setback, accident or delay.

By July 1946 the British Chiefs of Staff's Joint Technical Warfare Committee was estimating the potentials of biological and nuclear weapons. The conclusion was that, on a weight-for-weight basis, they were 'comparable' although BW's potential was admittedly 'speculative'. The authors, however, calculated a possible '100-fold increase in effectiveness' which, had it materialized, would have made biological weapons directly comparable with Nagasaki-sized nuclear bombs in terms of urban casualties. The report gave a rough idea of the current effectiveness of the anthrax weapon; 200 cluster units (21,200 bomblets), it said, would cause '50 per cent casualties to unprotected human beings in an area of about 3 square miles'. But despite the optimism, estimates on the effectiveness of the anthrax weapon seem to have actually worsened in light of further research. Two-hundred cluster units per 3 square miles is roughly the same as 67 per square mile and this is 7 more than the November 1945 report called for. Moreover, the 1946 estimate is for 50 per cent casualties compared with 50 per cent fatalities a few months earlier. In short, it would take more bombs to do less damage.

Five years later, a highly secret report on biological warfare to the American Secretary of Defense's Committee on Chemical, Biological and Radiological Warfare described the 1945 anthrax bomb as 'grossly inefficient' and said of BW research during the war: 'The work was necessarily of an applied nature and was based upon empirical assumptions for which there was often little factual justification.' The amount of anthrax carried in the 1945 bomblet amounted to only 0.75 per cent of its total weight, and of the total BW filling only 2 per cent was thought to be put into an effective cloud under ideal conditions. Dr Fildes reported in late 1945 that while anthrax was 300,000 times as toxic as the chemical war gas phosgene, the anthrax bomb was only '25 times as efficient in causing death'. 'In short', he concluded, 'phosgene is now distributed some 12,000 times more efficiently.' Despite some apparent improvements in anthrax's infectivity, estimates of bomblet efficiency actually declined by 1950. In that year a report gave the number of *improved* (E-96 cluster) 4-pound anthrax bomblets necessary to cause 50 per cent casualties as 11,420 per square mile. On the most generous assumptions, this represents a 43.7 per cent increase on the numbers given in the British document of November 1945 which would have supposedly caused 50 per cent fatalities.

Suspicion of Soviet work in biological warfare dates from that period. A post-war British report concluded that, apart from America and Britain, the only other country likely to have the capability to 'produce bacteriological weapons in effective quantities' was the Soviet Union. The report said, however, that 'there is no concrete evidence that she has done so' but that 'it is possible that the USSR has acquired information from the Japanese bacteriological weapons establishment at Harbin in Manchuria'. The 1950 American document hinted at Soviet biological work but no evidence was offered. After conceding that the West had been unable to develop a predictable epidemic-producing agent, the report warned of possible Soviet progress in this area by means of 'experimentation on human subjects'. Again, no evidence was cited but the report concluded: 'Certainly the controled subjects are available to him [the USSR], and it is likely that he would not have too much compunction to conduct experiments of this nature if he felt it worthwhile. The Nazi did it on concentration camp inmates. . . .'

The point about Nazi work is ambiguous. An earlier British BW report remarked that there was 'sufficient evidence' that the appalling experiments in German concentration camps had 'been confined to medical research and not bacteriological warfare'. The American claim seems to have been based more on the supposed efficiency of human experimentation than anything else. Noting that an actual appraisal of biological agents cannot be made 'short of their actual use in warfare', the report noted that the nearest 'approximation' might be large-scale field trials 'using human subjects', but that 'the War Council has decided we will not use such methods'. Since then there have been periodic reports of secret Soviet biological warfare installations quoting 'official' but unnamed sources. These installations are said to be on the shores of the Caspian Sea, in or near Moscow, in various locations throughout Siberia, the Urals and on the Iranian and Chinese borders. None of the reports have ever been publicly confirmed.

In the very first years after World War II biological warfare had three major applications in the eyes of the researchers: strategic, clandestine and economic. It was optimistically regarded by its supporters as a possible alternative to the atomic bomb and British documents suggested it could be used in small wars where neither side wanted to use nuclear weapons or in major conflicts where the nuclear option had been politically rejected. From the beginning biological warfare seemed impractical for the battlefield; the 1950 American report admits that the Chemical Corps had been unable to devise any tactical uses for biological weapons although it had hopes of doing so in the future. BW's clandestine possibilities remain a serious danger and are covered in a later chapter. Biological attacks on an enemy's economy could, it was thought, involve the use of anti-crop or anti-animal agents in support of strategic biological, nuclear or conventional warfare. In this, the proponents of biological warfare were basing their case more on the attitudes arising out of World War II than upon the thermonuclear age itself.

The original attraction of strategic biological warfare probably derived from the failure of the RAF's terror campaign against German cities in the last war. Although the bombing resulted in tremendous physical damage, its effect on civilian morale fell far short of Bomber Command's expectations. Strategic BW thus had two interrelated goals: to contribute to the elimination of a country's capability to wage war and to reduce the population's willingness to carry on the fight and its political support for the leadership. The latter is the psychological effect researchers referred to as the 'anxiety factor' — people's fear of disease and the 'invisibility' of micro-organisms. This, it was believed, would play a large part in breaking morale as well as overloading the health and emergency services, which would add to the spread of disease and cause further panic. Industrial productivity would be directly harmed as workers died, became ill or were drawn away to help treat casualties. It was thought that biological attacks could supplement chemical and conventional weapons in order to maximize the disruption to sanitary and health facilities, to open up buildings and shelters to invading micro-organisms, or to contaminate an area and hinder repair work on vital war industries. At the same time, attacks on animals and crops would initiate a longer-term decline in food resources, which would add the prospect of starvation to the immediate misery. There were other possibilities, too. A vital factory in friendly but occupied territory might be attacked with an incapacitating biological agent. The intention would be to render it ineffective by making the workers ill but not kill enough of them to antagonize the local population.

In the late 1940s and early 1950s when nuclear bombs were comparatively scarce and thermonuclear weapons were only being developed, biological warfare appeared a viable alternative but its attractions faded once the bomb became militarily practical and reasonably abundant. At the same time, the development of chemical nerve gases and potent incapacitants provided the military with weapons that were far more cost-effective and lacked the disadvantages of biological agents. Chemical agents do not have precisely predictable effects but they are much more reliable than biological weapons and are easier to store, generally cheaper to produce and more rapidly acting. In addition, chemical weapons have distinct tactical uses for the battlefield which biological agents do not. In any event, the idea of a war between NATO and the Warsaw Pact involving the mass bombardment of cities with non-nuclear weapons became more and more fanciful over the years. The moral distaste with which biological weapons were viewed by the general public, the politicians and the majority of the defence community contributed to the willingness to abandon them.

Apart from covert use, biological weapons are not likely to have any major military role in the immediate future. But 'defensive' research goes on. Genetic 'engineering' and manipulation offer an endless supply of new and devastating biological warfare agents but the important question is whether they could be adopted into practical weapons. On the battlefield, new BW agents would almost certainly have the basic disadvantages of the old ones. Protective field clothing and tank filtration systems designed to protect troops from chemical, biological and radiological hazards would be as effective against an 'engineered' micro-organism as a natural one. The position would be different if it became possible to manipulate epidemics or create a predictable and effective epidemic-producing agent. Such an agent would have little military value but might be regarded as an ultimate 'doomsday' deterrent.

The possibilities offered by the emerging techniques of genetic engineering seem open-ended. For a considerable time now, it has been possible to reproduce bacteria sexually, and cross-breeding two strains may result in a hybrid combining the desired qualities of both parents. A highly virulent, lung-specific strain might be bred with another extremely stable one, for example. Viruses may be effectively reconstructed. They consist of particular combinations of nucleic acid (RNA or DNA) in specific protein coatings and it is the protein not the acid that the body reacts to in its production of antibody defences. By deliberately altering the virus's protein, biological warfare researchers might be able partly to overcome immunity — natural or induced — in a target population. Diseases such as smallpox which have largely disappeared from the developed world could become viable and effective biological weapons.

Immunological studies aim at producing very low virulent strains through repeated cross-breeding and testing but BW work seeks to obtain increased virulence by the same procedures. One current goal of genetic engineering is to create a bacterium capable of producing large quantities of antibiotics. The idea is to take a class of bacteria known as streptomycetes which produce many antibiotics and transfer the responsible genes (plasmids) into a mutated form of *Escherichia coli* (the bacterial work-horse of contemporary genetics) which reproduces itself at several times the normal rate. These new bacteria would provide abundant supplies of antibiotics but would be dangerous to anyone ingesting them because they would attack the beneficial, symbiotic bacteria living in the human digestive tract. For safety, the hybrid organism would be specifically designed to survive only within artificial laboratory conditions.

Conceivably, however, it could have implications for biological warfare. First of all, the new bacteria could be converted into a weapon by breeding them for stability and survival outside the laboratory. Secondly, the idea might be adapted to the efficient production of toxins or in developing new or modified toxins. So far, the main reason why the techniques of genetic engineering have not been channelled into biological warfare research is because this work lies at the frontiers of knowledge and the truly qualified scientists are few and far between. As advances in genetic science progress, the specialized knowledge and techniques will become increasingly standardized and available to wider sections of the scientific community. When this happens, the possibility of monitoring the nature and direction of the work will become increasingly difficult and biological warfare laboratories will have a growing pool of expert knowledge on which to draw.

10 Chemical Weapons and Warfare

The Nerve Gases

Chemical weapons were not used in World War II, although various types were developed and considered. The major Allied discovery was chemical herbicides which were developed in Britain and handed over to the Americans for further work. By 1945 a plan had been drawn up to use these herbicides against Japan as part of the campaign to invade and conquer the Japanese home islands in 1945 and 1946. According to the plan, some 20,000 tons of the herbicide 2, 4-D (2, 4-dichlorophenoxyacetic acid) would have been spread across the 3.1 million hectares (7.8 million acres) planted in rice and distributed at around 2.25 kilograms per hectare (2lbs per acre). Two 2,4-D preparations were developed: the first was a 3 per cent solution in a mixture of tributyl-phosphate diesel oil, the second was a granular solid mixture for use against irrigated rice and was to be carried by a cluster munition. The liquid variety would have been carried by aircraft and distributed from spray or smoke tanks from altitudes of between 15 and 30 metres (50 to 100 feet). Also considered was an attack on Japan's 1.8 million hectares (4.4 million acres) given over to crops such as wheat and barley. This attack would probably have used isopropyl N-phenyl carbamate, also distributed at 2.25 kilograms per hectare.

The powerful herbicide 2, 4, 5-T (2, 4, 5-trichlorphenoxyacetic acid), a simple chemical extension of 2, 4-D, was also developed during the war but does not seem to have been included in the agents scheduled for use against Japan. After the war, however, 2, 4, 5-T became the constituent of many commercial weed killers and military herbicides including the infamous Agent Orange used in Vietnam. Many countries have now banned it as a commercial herbicide because of its dangerous side-effects which include the possibility of genetic damage.

Allied stocks of anti-personnel agents throughout the war consisted largely of agents developed during World War I. These included phosgene, the blood gases cyanogen chloride and hydrogen cyanide (also used in the United States for executions) and varieties of mustard gas. Phosgene corrodes the lungs and it has been alleged that it was used by the Egyptians in the Yemen. The two blood gases kill chiefly by blocking the enzyme in red blood cells that is responsible for removing carbon dioxide from the system. Although technically more lethal than the other gases, mustard gas is an oily liquid that is primarily used to inflict damage via the skin and eyes rather than to kill.

Had the number of available lethal agents not increased, the threat of chemical warfare (CW) may well have disappeared altogether, but the end of the war revealed several secret German compounds, first discovered in the late 1930s, which promised a quantum jump in toxicity. The least toxic of these gases, tabun (coded GA), is about 4 times as poisonous as mustard gas if inhaled and so chemical warfare for the battlefield received a new lease of life. These chemicals — which we have come to refer to collectively as nerve gas — are still the principal devices of chemical warfare and the chief obstacle to reaching agreement on a CW treaty.

Nerve gas originated in the insecticidal laboratories of the Wuppertal branch of I. G. Farben where Dr Gerhard Schrader first synthesized tabun, a highly toxic organophosphate. Tabun's military potential was recognized immediately and the gas was classified and taken over by the German Army. By 1939 Schrader was experimenting with organophosphates containing fluorine, and produced sarin (coded GB) which proved roughly four times as lethal as tabun. It was also quickly adopted by the German Army although the military never fully developed it. In 1944 Richard Kuhn synthesized soman (GD), the most toxic of the sarins. Known generally as the G-agents, the sarin chemicals had been successfully synthesized by 1945 by the British laboratories at Porton Down and the United States began to produce sarin in large quantities. Total American production of sarin is estimated at about 15,000 tons by 1957 at a cost of nearly $47 million to give a unit cost of between $1 and $2 a pound. (The comparative toxicities of the G-agents and other war gases are shown in notes 13 and 14, and their chemical names are given in note 15.)

About a milligram of sarin may prove fatal if inhaled. It may also kill through penetration of the skin but normally its rate of evaporation exceeds its rate of absorption. Sarin is highly volatile unless specially treated, and when it is dispersed it acts like a familiar riot gas, spreading out in a vaporous cloud. In its anti-personnel effects, nerve gas is directly comparable with kiloton-range nuclear devices and one of the standard explanations of the Soviet Union's build-up of chemical weapons in the early 1960s is based on the belief that the Kremlin was trying to find a counter to NATO's superiority in theatre nuclear weapons. Under 'ideal' weather conditions casualty-producing concentrations of sarin could drift more than 100 kilometres downwind. Thus there are many circumstances where sarin and the other nerve gases might prove a greater hazard to civilians in the country being defended than they would to an invading Red Army well supplied with anti-gas equipment and thoroughly trained in its use.

Sarin and the other major military nerve gases are chemically related to organophosphate insecticides such as malathion and parathion. Like these insecticides, the organophosphate nerve gases bind upon the vital enzyme acetylcholinesterate and inhibit neural transmissions by inactivating it. Death through asphyxia may follow within a minute of exposure to a large dosage and even small dosages may prove highly incapacitating, with death possibly following within a few minutes to an hour or so later. Symptoms are difficult breathing and then vomiting, muscle spasms, lack of control over bodily functions, possible paralysis, coma and death.

Efforts to secure an effective chemical arms treaty have in part foundered on an inability to make a clear scientific distinction between a viable organophosphate insecticide and an organophosphate chemical warfare agent. The only unambiguous difference that can be drawn is that the first is generally less toxic than the second, so that an inclusive treaty would have to ban chemicals over a certain level of toxicity rather than find an explicit definition of what was a war gas and what was not. There are also a number of non-organophosphate nerve gases which might be suitable for chemical weaponry and new candidate compounds may be discovered at any time. Thus an effective chemical arms limitation treaty depends not only upon meaningful inspection and verification clauses but also upon an exact agreement on what precisely is to be banned.

In the early 1950s chemical companies including Britain's ICI and Bayer (under Dr Schrader's team) began working with organophosphate esters involving substituted 2-aminoethanethiols. The result was amiton (VG), but previous experience with the G-agents caused researchers to investigate the toxicities of certain analogues. This work produced the V-agents and, of about a dozen, three became of immediate military interest in Britain and the United States. These were VE, VM and VX (for their chemical names see note 16). Only VX has actually been deployed as a chemical warfare agent and then only by the US Army, which produced about 5000 tons between 1961 and 1967. Unlike sarin, VX is non-volatile and highly persistent; it normally appears as a heavy oily liquid. It is more toxic than either sarin or soman and about half a milligram can be fatal when inhaled. It is much more effective if absorbed through the skin than sarin — although the percutaneous toxicity of any nerve gas can be improved by thickening it with synthetic polymers — and, depending upon the part of the body affected, as little as 1 milligram can be fatal upon contact.

As with sarin, the military effectiveness of VX depends upon several factors. Standard field clothing, for example, is said to offer a reasonable degree of protection against VX, more so in fact than against soman, which is less toxic when actually in contact with the skin. Weather is another important consideration and VX may persist for weeks in cold climates but only for a few days in warmer ones. When first produced, VX seems not to have been sufficiently stable for military deployment. The US Army's initial requirement for VX storage was that there should be no more than 20 per cent deterioration in steel containers at 71 °C over six months. But unless it is specially treated, VX may decay by that amount in just over three months. The problem was overcome by mixing it with stabilizing chemicals which reduced its deterioration to acceptable rates. This is important because toxicity is by no means the only crucial factor in selecting a war gas from the substances available.

There are many compounds more lethal than VX or sarin, for example, which were rejected because they were too unstable or for other reasons that made them unacceptable. This may be the case with the reported Swedish F-gas which is said to be more toxic than VX but has yet to be deployed anywhere in the world. For a time, the Soviets were believed to have deployed a V-agent known as VR-55 but this is now thought to be thickened soman — thickening of chemical agents reduces evaporation loss during dispersal, increases persistency and makes the gas more readily absorbed through the skin. An established

American method for thickening V-agents or G-agents is to mix them with synthetic polymers such as nitrocellulose or polystyrene. Between 0.1 and 2 per cent by weight of polymer is added to the CW agent and mixed for several hours. VR-55 is presumably produced in a similar manner.

Rival Capabilities

Stockpiles of chemical weapons are difficult to gauge because of the strict secrecy that surrounds the subject. This is especially true of the Soviet Union which officially admits to very little. Published Western estimates of Soviet chemical warfare stocks vary so widely that it is difficult to credit any of them. However, in *Chemical Weapons: Destruction and Conversion* Julian Perry Robinson gives what is probably the most realistic analysis of the chemical balance between East and West, when he suggests that the United States and the Soviet Union have roughly the same capabilities in offensive chemical weapons. He credits America with approximately 38,000 tons of casualty anti-personnel agents, just under half being nerve gas and the remainder consisting of three types of older mustard gas. Of the nerve gas about 75 per cent is sarin and the rest VX. About 65 per cent of the mustard and 25 per cent of the nerve gas are stored in bulk, and the rest is inserted into a variety of munitions — bombs, artillery shells, rocket warheads and land mines — amounting to between 150,000 and 200,000 tons. As no new weapons have been purchased since 1969, a high percentage of the American stockpile is now obsolete, dangerously leaking or unserviceable — the actual figure is unknown but as much as 40 per cent has been suggested. British stocks of nerve gas officially amount only to a few kilograms for research purposes. French stockpiles (most likely sarin) probably amount to a few hundred tons.

Estimates of Soviet stocks run as high as 700,000 tons or as much as 30 per cent of all munitions, but there is no evidence for these claims. What evidence there is suggests that the Russians possess a greater proportion of World War II chemical weapons' production than the United States — hydrogen cyanide and phosgene, for example. They almost certainly have stocks of soman, both in the thickened (VR-55) and unthickened form. The quantity is probably similar to American stocks of VX or marginally more. In addition, there are reports of tabun stocks remaining from captured German supplies and a nerve gas factory dismantled and transported back to the Soviet Union after the war. Chemical munition stocks are also thought to contain a fairly high proportion of older equipment — the 152mm artillery shell, for example, and a variety of bombs, bomblets and mortar shells. A high number of these munitions are probably filled with hydrogen cyanide which is held in bulk.

The Soviet Union is also thought to deploy chemical warheads — filled with hydrogen cyanide or thickened soman — for the short-range SS-1 Scud ballistic missile, the SS-N-3 Shaddock cruise missile and rockets such as the 112-kilometre (70-mile) Frog-7. These missile systems are frequently mentioned by those most concerned with Russia's alleged superiority in chemical warfare capability who draw a picture of Soviet missiles with chemical warheads striking

NATO airfields and incapacitating the pilots and crews if not actually killing them outright. For example, hydrogen cyanide may be severely incapacitating in minute dosages and is said to be generally more difficult to shield against by means of filtration systems.

There is no hard evidence that the Soviet Union possesses a chemical warfare capability significantly greater than NATO's or that it has continued any major development or production programme since President Nixon halted America's procurement of new chemical arms in 1969. In fact, chemical weaponry is one of the few areas in which no official, declassified estimates of Soviet capabilities have ever been published. Estimates are, however, made and most frequently quoted statistics is that the VKhV chemical corps within the Soviet Army consists of more than 100,000 troops. Although many have interpreted this force as evidence of a threat, many others believe it to be nothing more than a 'janitorial' service designed to decontaminate battlefields and equipment rapidly from all chemical, biological and radiological hazards. The extent to which America and Britain are seriously committed to chemical rearmament is difficult to judge. In 1980 Congress voted $3,150,000 for the design and construction of a new nerve gas plant at Pine Bluff, Arkansas, and the Senate has since approved around $20 million for equipment. By the summer of 1981 the final decision to proceed with nerve gas production had not been given but was expected following the completion of an Anglo-American study on the value of chemical warfare. On the other side of the Atlantic, shortly before leaving office in 1981, the Secretary of State for Defence, Francis Pym, hinted that Britain might possibly readopt chemical arms and implied that the government was giving thought to producing them in the United Kingdom.

However, it may be that these moves are primarily intended to demonstrate military resolve to Moscow and to persuade the Soviet Union to re-enter serious negotiations on chemical disarmament. It is clear, nevertheless, that there has been an increasingly vocal lobby pushing for new chemical arms in both Washington and London. The actual military value of chemical weaponry will be discussed later, but it has never been popular in the military community and, to a large extent, it runs contrary to current strategies of reduced collateral damage. But if production does begin at the Pine Bluff plant, it will not be simply for nerve gas but rather for two individual chemicals which will make a nerve gas upon mixing in a weapon system known as a binary.

Binary Chemical Weapons

In 1968 some 6000 sheep grazing near the US Army's Dugway Proving Ground, Utah, were accidentally killed during an open-air test of VX some 45 kilometres (28 miles) away. The outcry which followed is thought to have been one of the chief reasons behind the decision to ban the purchase of new chemical arms the following year. At the time of President Nixon's order the army was developing a new range of chemical weaponry known as binaries as an answer to the dangers of storing and transporting nerve gas. Now that the 1969 ban looks like being reversed, proponents of chemical rearmament regard the binary system as a means of appeasing environmentalists and convincing Congress that the chances of accidental release have been reduced to a minimum. The sole advantage of the binary system is this safety feature but it is militarily less efficient than what it is designed to replace.

The idea behind the binary is to create the toxic agent at the point of use rather than manufacture it for long-term storage or for immediate insertion into a delivery device which is then stockpiled until needed. To do this, the binary weapon is designed to have two separable units, each of which contains a different and relatively non-toxic chemical. One unit will be stored and transported separately from the weapon itself until such time as it is to be used. At this point the second unit will be inserted into the weapon and, upon firing, the two chemical precursors will be brought together to meet and mix. In theory serious accidents are ruled out because the lethal product does not exist until the final seconds before its detonation. Binary designs have now been effected for virtually all the previous generation of chemical weapons and precursors have been developed for at least VX and sarin.

The favoured American approach for VX appears to be a reaction between liquid QL (ethyl 2-diisopropylaminoethyl methylphosphonite) and NM (a dimethyl polysulphide) in a powdered form. For sarin the present binary route consists of reacting DF (methylphosphonyl difluoride) with IP (isopropyl alcohol). DF is a slower reactant than other potential precursors but is preferable because its chief waste product, hydrogen fluoride, has a lower molecular weight than other binary sarin by-products. This means that there is a higher proportion of sarin in the final mixture. To accelerate the reaction, a promoting agent is added to one of the precursors. The reaction time is about 10 seconds and the proportion of sarin in the final mixture is approximately 70 per cent. One US Army document refers to a binary VX route with a reaction time lasting three to five seconds but it does not specify whether or not this involves the precursors QL and NM. There are other candidate precursors for both VX and sarin but the two routes described are those most commonly mentioned.

The first of the new generation of chemical weapons will be a 155mm artillery shell which, because it exploits the physics of inertia and spin, is ideally suited to the binary mode. Two canisters are inserted end-to-end behind a burster charge in the nose. When the shell is fired, inertia breaks the rupture discs separating the forward canister containing DF from the rear canister containing IP and forces the DF back to meet and react with the IP. At the same time, the spin imparted to the shell by the barrel's rifling ensures a rapid and complete mix. A 203mm shell for VX is also under development and, further back in the pipeline, extended-range, sub-calibre 203mm binary projectiles for both VX and sarin.

In the older, non-binary chemical shell, the burster charge runs through its length and is surrounded by a chamber containing finished nerve gas. The older design creates a shrapnel hazard in addition to a chemical one because the burster charge ruptures the entire casing into tiny fragments. The binary design, on the other hand, will merely break the shell into a few pieces. The older

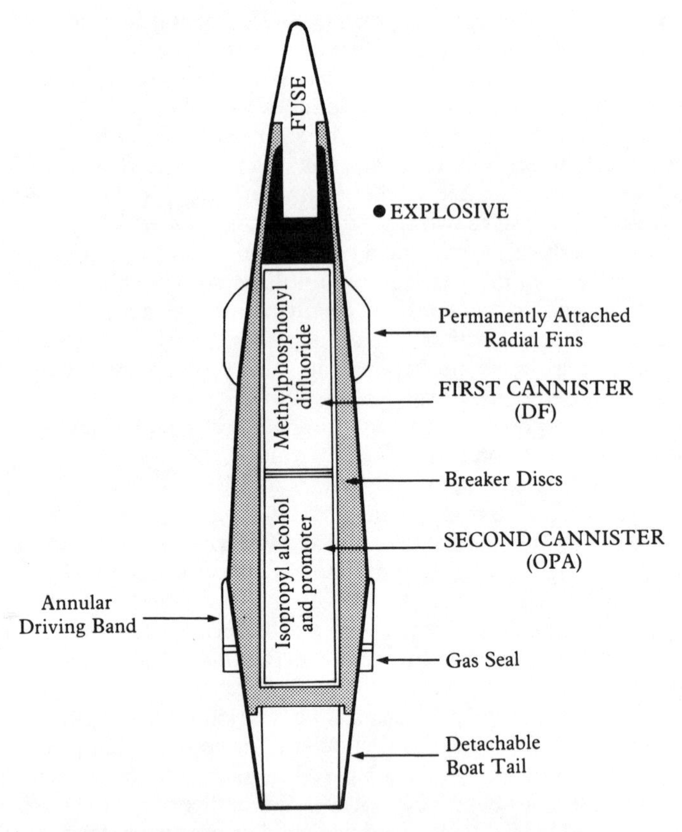

Fig. 17 Extended-Range Binary Nerve Gas Artillery Shell

The shell as drawn is an extended-range design by Space Research Inc and reportedly purchased by South Africa as a nuclear delivery vehicle. Such shells may be equiped with chemical, nuclear or conventional explosive payloads.

design also disperses the chemical agent in more-or-less even directions unlike the binary which will restrict the area covered per shell. An extensive series of open-air tests of live munitions will be necessary to see how these shells actually behave in practice under different environmental conditions. The use of simulants — non-toxic chemicals which mimic many of the important characteristics of nerve gas — will reduce the need for live ammunition but it will still be necessary to create binary nerve gas under simulated battlefield conditions.

The second binary weapon scheduled for production is the 225-kilogram (500-pound) Bigeye VX bomb. Unlike artillery shells, binary chemical bombs cannot utilize inertia and spin in the same efficient manner and consequently require a fairly complicated set of internal gadgetry to bring about the final

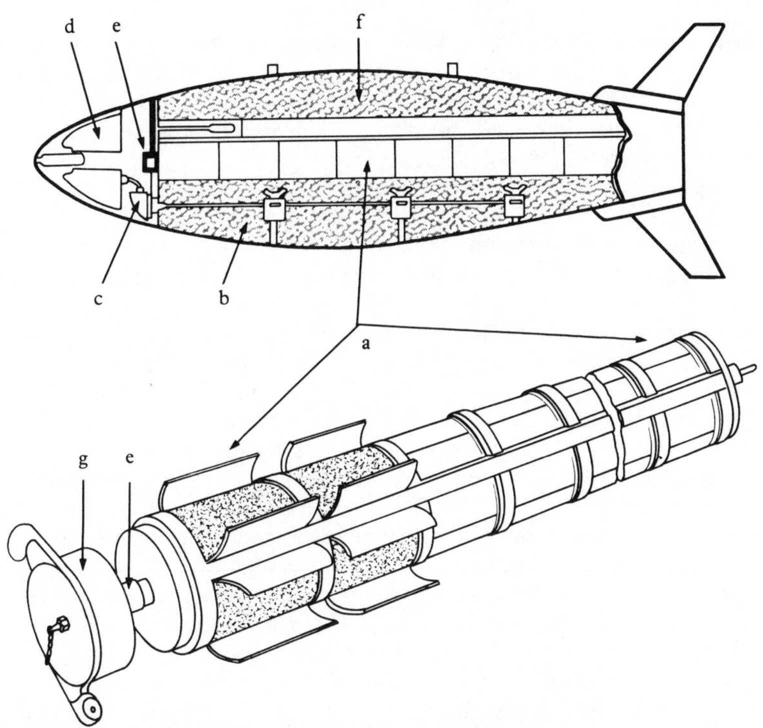

Fig. 18 Basic Outline of Design for Binary Bigeye VX Bomb

When the pilot ignites an explosive charge in the chamber (e), the resulting gases are forced down the multibar injector device (a) which contains the powdered sulphur (shaded area), breaking the partitions on the outer wall and forcing the chemical into the surrounding chamber (f) containing liquid QL. As the two chemicals meet, the VX reaction begins.

The pilot then accelerates the reaction by either starting a motor (c) powered by a tank of pressurized helium (d) which turns a shaft (b) and activates the propellers linked to it for a rapid mixing action, or the pilot activates a rocket motor (g) attached to the end of the explosive chamber (e) which then rapidly rotates the multibar device. The partitions of the multibar's wall — now broken outwards by the force of the initial explosion — act as paddles and complete the VX mix. (From US patent 3691952.)

reaction. The above illustration shows a design for Bigeye and the added complexities are immediately apparent. The bomb contains a mechanism for bringing the two precursors together and additional gadgetry for mixing them completely. In effect, the binary mode has imposed a number of design complications on a simpler system. Older nerve gas bombs are little more than exploding tanks of lethal liquids but binaries will contain a number of essential mechanical elements which will increase the chances of malfunction and therefore the unreliability of the weapon. With certain liquid-solid binary combinations, as in the QL-NM route to VX, the product may begin to deteriorate after reaching its maximum proportion because of heat generated within the munition. Thus there may be severe constraints upon both the lower and upper limits for delivery times and, for aircraft-delivered bombs, restricted operational patterns in combat.

Binary chemical munitions are probably the only major example of the military deliberately opting to replace an efficient system with one that will be less reliable and more restricted in application. The sole advantage is the binary's greater safety, but even the precursor chemicals may not be as harmless as the Pentagon maintains. The sarin precursor DF, for example, can be poisonous in its own right and as Julian Perry Robinson has written: 'Only Chemical Corps people, used to the extreme hazards of nerve gas, could describe it as "harmless" or "relatively non-toxic".'

Other binary weapons under active consideration for future production include warheads for a number of tactical and theatre missiles including the short-range surface-to-surface Lance and, even the cruise 2400-kilometre (1500-mile) Tomahawk. The design and engineering complexities of a binary missile warhead rival those of the Bigeye bomb and Lance is unlikely to be a cost-effective chemical delivery system when compared with other payloads. The same is true for Tomahawk and other cruise missiles; the alternative payloads are simply of greater military value. When Tomahawk's accuracy is taken into account, nerve-gas cruise missiles seem even more implausible because there is little need to deliver chemicals with such precision and at a cost of well over $9 million per missile. Like any other specialized military group, the Chemical Corps plans for delivery systems in excess of what would be realistic so that Lance or even Tomahawk will probably have a binary chemical warhead designed for them. More practical possibilities include binary air-to-ground rockets and 81mm binary mortar shells. Chemical munitions designed to smash their way through concrete walls and armour plating have been developed but it is doubtful if these could be adapted for binary payloads because of the mechanical complexities involved. The illustration on page 135 shows a TOW missile chemical warhead capable of penetrating 25 centimetres (10 inches) of armour and dispersing a lethal or incapacitating agent. Similar designs have been completed for shells and bombs but any future deployments are likely to carry non-binary payloads.

A third type of binary nerve gas may join VX and sarin. For several years the Chemical Corps has been interested in developing an intermediate volatility munition (IVM) armed with an agent possessing dispersal and persistency characteristics midway between VX and sarin. Soman was investigated but was set aside. Another candidate is still known only by its Edgwood Arsenal code number, EA-5365, and may be an entirely new nerve agent or an older one in modified form. A binary IVM munition might be based upon the production of thickened sarin, produced by blending the thickening agent with one of the sarin precursor canisters. A commander could then choose the level of volatility he wished by making a selection from differently coded canisters for insertion

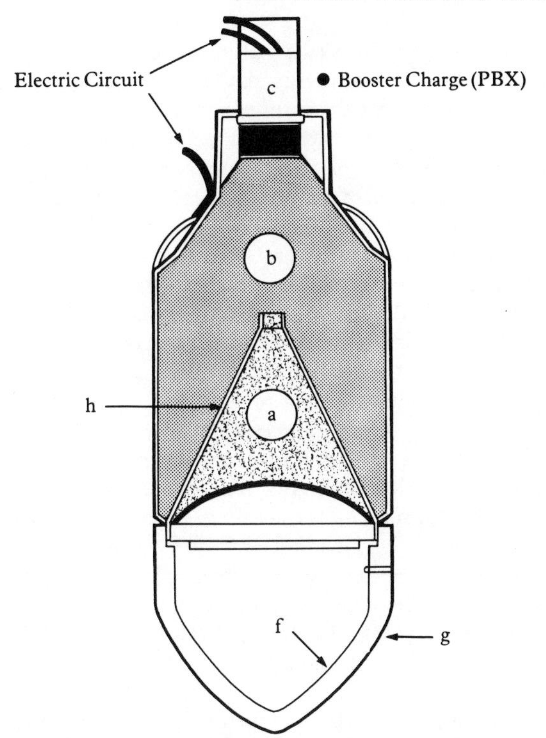

Fig. 19 Penetrating Chemical Warhead (shaped-charge TOW missile)

About 70 metres after firing, the arming device (c) activates. Upon striking the target, the outer nose cone (g) is crushed into contact with the inner cone (f) completing an electric circuit causing the arming device (c) to fire the PBX booster charge. The booster ignites the shaped-charge explosive (b) which directs its force forward to bore and cut its way through the target. At the same time, the aluminium cone (h) is crushed and the toxic chemical agent (a) is drawn into the target by the explosive forces ahead.

The drawing shows the front portion of a TOW missile but the technique can be adapted to shells and aircraft gravity bombs.

into the binary sarin shell. It might be even better to find a chemical which acted both as a thickening and as a skin transferral agent. When certain chemicals (NN-dimethylpalmit-amide and dimethylsulphoxide, for example) are mixed with a nerve gas, they increase its rate of action through the skin and thereby reduce the time to death or the amount of agent required.

One definite advantage of the binary approach is that it increases the number of agents which might be suitable for chemical warfare. Literally hundreds of highly toxic chemicals have been investigated over the years but only those described here are known to have been deployed. Many cannot be made sufficiently stable to make them serious contenders but, in a binary system, it would only be necessary to find two stable precursors which made the end-product efficiently when they were mixed. One such class of compounds which attracted considerable interest for a time was declassified in 1977. They are generally known as O-alkyl and O-cycloalkyl methyl phosphonofluoridothioates (see note 16). But they were finally rejected; why is not certain. Another highly toxic chemical of passing interest but of insufficient stability for at least non-binary deployment is homocholine (see note 17) but it does not ever seem to have been a serious contender.

Trying to predict new binary or non-binary chemical warfare agents that might be adopted is difficult as literally hundreds of candidates arise each year. Most agents studied are first discovered in the course of industrial or general scientific research and are then examined in detail in specialized defence laboratories where more suitable variations may be synthesized for military application. The quaternary compounds, for example, have long been known but their toxicities and mechanisms of action apparently vary widely. The US Army's work was devoted to exploring highly toxic varieties which disrupt neuromuscular impulses by mimicking the action of acetylcholine. As far as toxicity is concerned, the work was a clear success; one compound (note 18) showed itself to be about twice as poisonous to rabbits as VX and roughly the same as VX when injected into mice. Another (note 19) had a VX-like toxicity to rabbits but was at least 1.3 times as lethal to mice. On this basis, the human LD_{50} dose is in the order of 0.22 milligrams compared to a minimum of 0.35 milligrams for VX.

As a general guideline, new lethal agents would almost certainly have to fulfil the following minimal conditions. First, they would probably have to be extremely toxic. Secondly, in accordance with current trends, be suitable for binary weapons. This would mean that if the chemical itself were insufficiently stable for military use, its two precursors would need to be adequately stable — either on their own or with the addition of stabilizing chemicals which do not detract from the binary reaction. Thirdly, any new war gas would require a high degree of percutaneous toxicity either of itself or through modification with skin transferral agents. The binary precursors will complete their reaction within a few seconds and produce a very large proportion of the final product which should be stable for a reasonable time afterwards.

There are reports that the Russians are interested in binary research, despite the fact that they are not burdened with a vocal environmentalist lobby. Their interest arises chiefly from the possibility of finding stable precursors for agents that would otherwise be unsuitable, and they are also attracted by the increased safety features of binaries. But, as yet, there is no evidence that this research reflects any serious commitment to the development of binary munitions. Because binaries appear to have too many military disadvantages compared with older chemical weapons the Soviet Union has no obvious reason to adopt them. If the United States proceeds with a full-scale binary programme, however, the Russians will almost certainly respond with a renewed stockpiling of chemical arms. For the immediate future this will probably involve the replacement of older stocks of mustard gas and phosgene with newer agents such as thickened soman and possibly a new but well-understood nerve gas such as a V-agent or sarin.

Any American commitment to chemical rearmament could well have a decided influence on smaller powers. The large-scale manufacture of nerve gas requires a considerable industrial base and expertise but binaries offer both a simpler method of manufacture and one which eliminates many of the necessary complex and expensive safety arrangements. Generally, making nerve gas in large quantities requires an advanced and sophisticated civilian chemical industry which does not exist outside the major industrial powers. With binaries, on the other hand, it is only necessary to prepare the precursors – which is neither involved nor dangerous. Some binary precursors (isopropyl alcohol, for example) can be purchased commercially and there are products now on the market which can be converted into DF fairly easily and reasonably safely. Nor would a small power be deterred by the technicalities of American binary munitions. Instead of mixing the two precursors within the munition at the point of use, it would be perfectly feasible to prepare the nerve agent in advance and then insert it into a simpler weapon of the older type. The procedure would be more dangerous but it would not inhibit a minor power anxious to have a chemical warfare capability.

The Incapacitants

Since the 1950s Porton Down and the American Chemical Corps have investigated the possibility of 'bloodless warfare' by chemicals which would incapacitate without killing. These chemicals are quite distinct from harassing agents such as tear or vomiting gases where the effects are brief; true military incapacitants may be effective for several hours or even days. Although the applications are intended to be the same, the chemical incapacitants also differ from biological incapacitants scientifically as well as in the effect they have on victims. Biological incapacitants, such as US, may affect an individual for a long time and, if used on any large scale, would involve a small proportion of deaths among the physically vulnerable or those who suffered further complications following the initial infection. However, chemical incapacitants are intended to kill in only very unusual circumstances and to take effect within minutes or hours compared to days in the case of biological agents. There are two approaches, physical or mental incapacitation, but both carry problems which have still to be fully solved.

For physical incapacitation, the problem has been to find agents which caused a disabling symptom such as muscle paralysis, interference with the voluntary control of the limbs, unconsciousness or loss of sight, but would be very unlikely to cause death. Those which affected neural transmissions between muscles offered the best likelihood of incapacitation but are also the most dangerous. Neuromuscular inhibitors may not be sufficiently selective and can kill by inhibiting the muscles which control breathing, for example. Two sets of physiological incapacitants that were briefly investigated – the haloalkyl-carbamoxyalkyl derivatives and quaternary quinuclidinones —added decreased respiratory rates and tremors to decreased locomotor activity and sensitivity to touch. But these compounds have safety ratios of only 10 and 40 respectively. Others were even more dangerous; two neuromuscular inhibitors, for example, have safety factors of only 1.8 and 3.1.

In practical military terms, no incapacitant with a safety ratio of less than 100 could be seriously considered for deployment on a large scale, although several with less stringent safeguards have been stockpiled in small quantities for special operations and even riot control. Toxin incapacitants which disable through severe diarrhoea and vomiting were developed by extracting the active constituents of bacteria such as salmonella and staphylococcus, but the one standardized toxin incapacitant, PG, is now banned by the terms of the 1972 Biological Weapons Convention.

For a long time, psychological incapacitants seemed to offer a more fruitful approach because they had far higher safety ratios, although many do have some physical effects, causing eye dilation, vomiting, dizziness and headache, as well as mental disorientation. In the early 1950s universities and pharmaceutical companies in the United States and Britain began receiving government contracts to carry out research on various psychotomimetic chemicals. (Terms with much the same meaning are 'psychotogen', 'psychochemical' and, to a large extent, 'hallucinogen'.) Porton Down's D. F. Downing defined the psychotomimetics as chemicals which 'constantly produce changes in thought, perception and mood' without causing 'major disturbances of the autonomic nervous system'. There are certain types of mental disorder such as schizophrenia which specific drugs will mimic for short periods. Research suggested that the psychotomimetics interfere with three bodily compounds vital to the normal functioning of the brain: 5-hydroxytryptamine (serotonin), adrenaline and acetylcholine. Blockage of acetylcholine was thought to inhibit the parasympathetic (depressant) centre in the brain allowing the excitatory centre to dominate.

Among the first psychotomimetics investigated were several homologues and analogues of the chief active constituent of *Cannabis sativa*, delta-one-tetrahydrocannabinol (THC), such as DMHP, which is more than eight times as effective as THC itself. Other highly effective nitrogen analogues of THC were synthesized in the late 1960s and at least one (an azatetrahydrocannabinol) is some 16 to 17 times as potent as THC. As incapacitants their purpose would have been to reduce military effectiveness through induced lethargy and impaired motivation rather than extreme mental disorientation. Cannabinoid agents also incapacitate through hypotension — imbalances in blood pressure that can lead to temporary unconsciousness. Another common hallucinogen which received a great deal of attention was mescaline, the principal active ingredient in the peyote cactus. Mescaline (3, 4, 5-trimethoxyphenethylamine) has several chemical analogues which have been synthesized and were also investigated in defence laboratories. Some of these analogues produce hallucinogenic effects in smaller dosages than mescaline itself (see note 13).

All the world's well-known hallucinogens were examined including those found in plants well known in folklore. Scopolomine, for example, is the active ingredient of several psychoactive plants including *Hyoscyamus niger* which is common to many warm climates and has been used as a poison and in demonology since the Middle Ages. Also examined were psilocybine, psilocine, *d*-lysergic acid, bufotenine, nalorphine and ibogaine. The first two are the active ingredients of a number of hallucinogenic plants including the sacred mushrooms of southern Mexico and *d*-lysergic acid is found both in morning glory seeds (which had a long tradition of religious use in Aztec religion) and in ergot, a rye-infesting fungus found in cereal-producing countries. Ergot has a long history of medicinal and hallucinogenic use in Europe and is the natural source of ergotamine, from which LSD (lysergic acid diethylamide) is chemically prepared. Ibogaine has cocaine-like effects and is found in the roots of a west African shrub. Bufotenine is a hallucinogenic found in several species of toad and fungi while nalorphine (N-allymorphine) is the synthetic analogue of a morphine alkaloid found in the opium poppy. Unlike most of the morphine alkaloids or their analogues, nalorphine produces hallucinogenic effects.

Purely synthetic chemicals with no natural source attracted attention and some became drugs of abuse once the knowledge of their psychotomimetic properties and formulae escaped the confines of defence laboratories. First supplies of the street-drug STP (2, 5-dimethoxy-4-methylamp' etamine) are reported to have come from a chemical warfare laboratory but it may have only been specific information on the drug's effects and the details for making it that emerged from this source. LBJ (N-methyl-3-piperidyl benzilate hydrochloride), now another reasonably common street-drug, originated with investigative work at America's Lakeside Laboratories, Wisconsin, and the US Army's Edgwood Arsenal during research into an extremely potent class of psychoactive chemicals known as glycolates. Although the first work on the glycolates occurred in the late 1950s, it took about a decade for LBJ to make the transition from the research laboratory to the street; STP, by contrast, took only a year or two. The difference was in no small part due to the growth in drug consciousness that had occurred in the United States in the intervening years.

One of the most commonly investigated candidate incapacitants was LSD as well as its more potent analogues such as MLD-41 and ALD-52 which is of equal potency. LSD is the most powerful hallucinogen known; as little as 2 micrograms per kilogram of body weight can induce hallucinations and mental disorientation for one to six hours or more. The equivalent dosage for mescaline, by contrast, is about 5 milligrams per kilogram, some 2500 times the minimum LSD dose. But LSD appears to have been rejected because of the comparatively high cost of large-scale production and suspicion that its use might cause long-term genetic damage. The Russians seem to have come to the same general conclusion because they are not known to have ever deployed LSD, despite intensive work on the drug. However, during the investigations leading up to Britain's Operation Julie LSD arrests in 1977, Detective

Inspector Richard Lee learned that a British company had sold 400 million doses of LSD to the People's Republic of China, possibly for use in chemical warfare.

The main incapacitant to be produced and deployed in chemical weapons is BZ (3-quinuclidinyl benzilate). It was first developed by Hoffmann-la-Roche in the early 1950s. BZ is a member of the glycolate family and belongs to a subgroup of powerful psychoactive chemicals referred to as benzilates. Between 1963 and 1964 the US Army manufactured approximately 45,000 kilograms (100,000 pounds) of BZ for bulk storage or insertion into weapons. The two principal BZ weapons were the 340-kilogram (750-pound) M43 bomb cluster and the 80-kilogram (175-pound) M44 generator cluster munition. The M43 carries 57 4.5-kilogram (10-pound) M138 bomblets filled with about 680 grams (24 ounces) of BZ; the M44 has three separate BZ canisters. The payload usually consisted of 50 per cent BZ with the remainder being a mixture of pyrotechnic compositions which, upon ignition, dispensed the agent in a dense smoke. Mixtures included BZ with potassium chlorate and sulphur blended with a polyurethane binder, or else BZ combined with potassium chlorate (25 per cent), sodium bicarbonate (15 per cent) and sulphur (10 per cent).

BZ is an extremely potent psychoactive chemical and miniscule amounts may produce effects lasting 90 hours or more. The minimum effective injected dose of LSD is only marginally smaller than BZ's, but the effective oral LSD dosage is around 15 times less. Symptoms of exposure to BZ may include vomiting, stupor and a lack of co-ordination and occur within an hour or so. After this, mental disorientation and hallucinations (both audio and visual) may incapacitate the victim — sometimes to the point of immobility — and maniacal behaviour may follow. Severe behavioural disturbances may last or reappear sporadically for the following two or three days until the drug finally wears off. Partial amnesia is a further effect.

Ironically entitled *Trip Report*, a 1975 CIA summary of another potent benzilate, EA-3167, reported on tests on volunteer convicts and military personnel. Delirium and other psychotic behaviour lasted for three to four days with subsequent amnesia. Some residual effects lingered for up to six weeks. Mental incapacitation involved visual and auditory hallucinations, an inability to relate to surroundings, time distortion, degraded performance on numerical tests and forgetfulness of such things as names. Hallucinations included conversations with imaginary people, complaints of bright lights and non-existent insects flying around the room. Although benzilates such as BZ or EA-3167 are less potent than LSD in terms of effective dosages, they are generally more powerful as far as the duration of symptoms is concerned.

BZ has now been phased out of the American chemical arsenal. Numerous studies of BZ, LSD and other psychoactive compounds on volunteers (and sometimes unsuspecting servicemen) in Britain and the United States failed to establish a sufficiently predictable pattern of military ineffectiveness due to psychoactive incapacitation. Indeed, motivation might actually be enhanced under mental disorientation rather than decreased. In battle, these agents would have been dispersed in the same manner as lethal ones over a wide area by a variety of cluster bombs and generators in order to render the enemy ineffective and passive. It would be a relatively humane but still effective form of warfare.

Incapacitating chemical warfare could be used in allied or friendly territory and, if winds did carry the agent into civilian areas, the victim would suffer only the effects of a psychoactive drug rather than death or severe illness. But none of this is attractive unless the weapon will affect the enemy in the way it is supposed to — and the psychochemicals cannot guarantee this. Furthermore, an enemy is not likely to consider defeat by 'bloodless' weapons more 'acceptable' than defeat by lethal ones. On the battlefield, therefore, incapacitants would be no less potentially escalatory than nerve gas or any other lethal anti-personnel weapon.

There are specialized operations in which effective incapacitants would be invaluable but they would rarely be straightforward military engagements. Probably the best example is President Carter's ill-fated attempt to rescue the American hostages held in the Tehran embassy in 1980. According to several press reports at the time, had it succeeded the raid would have involved spraying the embassy compound with an incapacitating chemical — perhaps a benzilate mixed with a more rapidly acting type of 'stun' gas. Several such incapacitating gases have been developed that can quickly overcome terrorists or rioters; in June 1980, for example, the West Germans announced the development of a riot gas which would immobilize demonstrators for up to half an hour without causing harmful side-effects or making them unconscious.

Recent military research into incapacitating chemicals has concentrated on physiological rather than psychoactive agents such as the benzilates EA-3167 and EA-3834, which were developed as potential successors to BZ. Instead of pursuing chemicals which affect the entire autonomic nervous system and carry a large risk of killing, researchers have investigated agents whose action would be specific to certain parts of the body, such as the nerve ganglia in the legs or brain. The victim would thus only be immobilized. These incapacitating gases would be put into a range of gas grenades and canisters suitable for use by both police and troops as CS is today. Without a serious commitment to lethal agents, however, extensive deployment of incapacitants is unlikely since any decision to adopt non-lethal gas weapons on a major scale could be interpreted as an intention to embark upon full-scale chemical rearmament.

The Value of Chemical Weapons

Western chemical strategy is generally based upon VX with its persistence and sarin for its threat to unprotected troops. Sarin dissipates quickly but VX may contaminate an environment for weeks, creating a long-term hazard. VX has a distinct role in direct anti-personnel attacks but its chief advantages over sarin are its greater percutaneous toxicity and its persistency. Used together, sarin and VX could create both immediate and longer-term hazards.

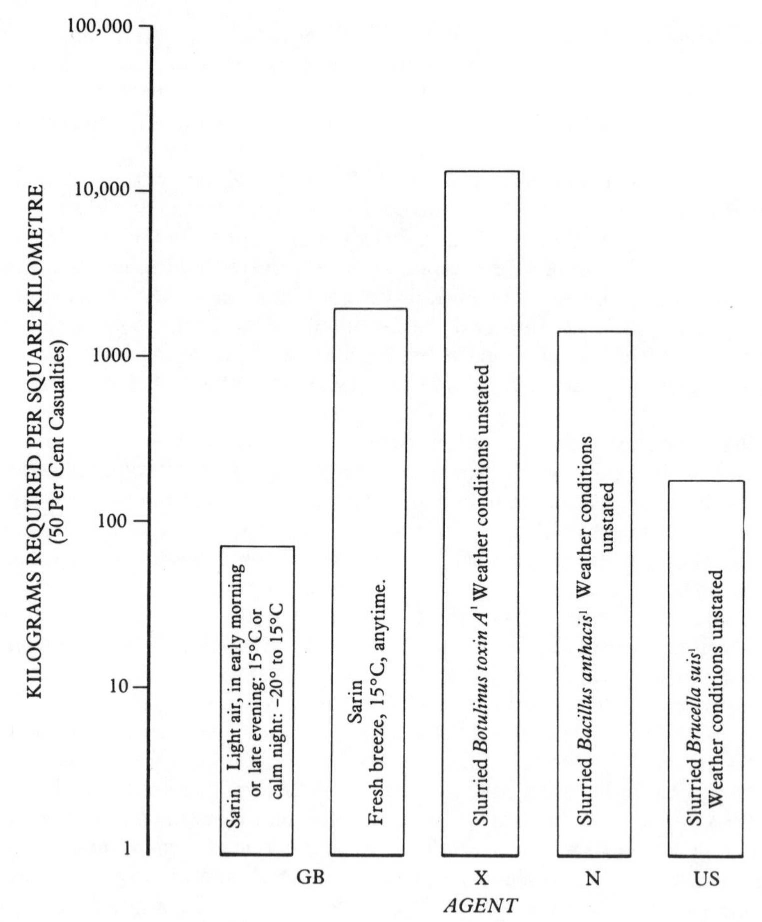

Fig. 20 Chemical and Biological Warfare Effectiveness

The graph above shows an approximate comparison of the effectiveness of sarin and three of the older biological agents in the E-96 cluster 4-pound bomblets. Because of differences in data and parameters, the figures are not directly comparable but, taking everything into account, the graph does give a reasonable idea of the relative effectiveness of CBW agents.

Although nerve gas has never been specially designed for the strategic bombardment of urban areas, it could be highly effective. In 1975 the US House of Representatives Armed Forces Committee heard testimony that a single 500-pound sarin bomb could — under certain weather conditions — cause 50 per cent deaths over an area of 1.3 to 5 square kilometres (0.5 to 2 square miles) if exploded in Washington D. C. Casualties would be extensive over an even wider area.

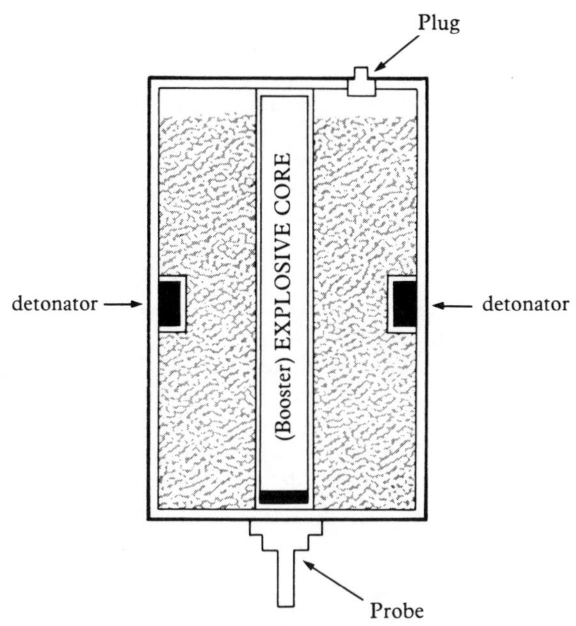

Fig. 21 Fuel-Air Bomblet

But the tactical advantages of chemical weapons can be duplicated or exceeded by improved conventional munitions. Fragmentation artillery shells, for example, may prove far more effective against troops in the open which are well prepared in chemical defences. Potentially far more effective yet in battle are fuel-air munitions. Although the subject of experiments since World War II, the fuel-air explosive (FAE) is only now being fully exploited in the West and, according to reports, the Soviet Union. FAE may be inserted into free-fall bombs, bomblets (for cluster munitions), artillery or mortar shells, or rocket warheads. It is a liquid which forms a highly volatile vapour cloud when the weapon detonates and forces it outwards to mix with the air. The size of the cloud depends upon the size of the weapon and may range from a few square metres to several tens of square metres. Along with the FAE liquid, the weapon ejects one or more detonators and after a pre-set delay they fire and ignite the cloud creating a downward blast with a force of as much as several hundred pounds per square inch.

These blast pressures are directly comparable with nuclear weapons but the damage done by FAE is for the most part limited to the area formed by the cloud and involves little or no collateral damage. Current uses envisaged range from clearing foliage and minefields by saturating the area with small FAE devices to attacking fortified troops — a role in which FAE would be far more effective than chemical agents. Another possibility is the FAE land 'mine': a set area is sprayed with the fuel and, when enemy troops and vehicles enter, it is detonated making the ground erupt in a violent explosion. FAE are said to be virtually undetectable by smell and, once treated, the ground remains effective for more than a day — longer-lasting fuels are under development. Such an approach could be a far more effective and predictable area-denial technique than persistent chemicals.

A cylindrical FAE bomblet designed to carry 32 kilograms (70 pounds) of ethylene oxide is illustrated on page 143. The bomblet is not totally filled so as to allow the fuel room to expand during storage. A parachute is attached to the rear to enable it to hit the ground at a selected speed. When it does, a probe at the nose triggers a fuse and sets off a booster charge at the front of an explosive core running through the bomblet. The fuel is forced outwards to mix with the air and form a 3-metre thick cloud with a radius of about 6 metres and is detonated 125 milliseconds later. The final proportion of fuel to air is about 6 per cent with lesser amounts of fuel in the centre and larger amounts on or near the circumference. FAE has weaknesses in that the cloud is vulnerable to heavy winds and in what is known as the 'detonation limits' of the fuel — the proportion of fuel to air which determines whether or not it will explode uniformly or simply burn.

Nerve gas may appear to offer a partial counter to the fluid pattern of warfare some see arising from improved conventional munitions. In a toxic environment front-line troops would have to be provided with temporary rest and aid stations to accommodate the injured, allow for gas mask and anti-chemical clothing changes and other requirements such as bodily functions. Quite simply, chemical warfare would impose a degree of fixed-place deployment and reduce the potential mobility of troops; in effect, driving troops onto the killing ground where they could be more easily targeted.

At present, the sole justification for chemical rearmament in the West is to deter the Soviet Union. Most supporters of new chemical arms argue that NATO's present weakness lowers the nuclear threshold. Faced with reversals from the Soviet use of chemicals, they argue, NATO commanders would have little choice but to call for the early use of theatre nuclear weapons and, therefore, it is preferable to have a chemical deterrent. This argument presumes that the Soviet Union would not wish to postpone a nuclear decision itself. If Moscow preferred to avoid using nuclear weapons, it would not be likely to quickly resort to chemicals if that were thought to lower the nuclear threshold. In fact, NATO's chemical weakness might effectively raise the CW threshold as far as the Soviet Union is concerned. A parity in chemical weapons, however, might make their use appear more feasible without the risk of escalation into nuclear war. Most proponents of chemical rearmament overlook the simple fact that Russia's so-called 'massive' chemical arsenal — if it exists at all — is probably directed more at China than at NATO. China is not so well equipped nor so well trained in anti-gas defence as NATO and it was partly America's experience of Chinese 'human wave' attacks in Korea that spurred on interest in the potentials of nerve gas in the early 1950s.

11 Covert and Terrorist Uses of Chemical, Biological and Nuclear Warfare

Security and Intelligence

The clandestine possibilities of chemical and biological warfare became of interest to the CIA and specialized military groups in the years following World War II, and the US Army's investigations of covert applications appear to have been largely taken over by the CIA by the middle 1950s. In 1954 a research project into chemical, biological and radiological (CBR) agents was created within a highly secret programme known as MKULTRA, a vast scheme investigating such diverse topics as mind control and extra-sensory perception. It was funded from a special budget that was not subjected to normal government or CIA accountancy procedures. By the late 1960s MKULTRA had been incorporated into a new programme known as MKSEARCH and among its numerous projects were at least two involving chemical and biological agents. Work under MKSEARCH continued long after President Nixon's order to destroy all stocks of biological agents, and later investigations by the US Senate showed that the CIA had secretly maintained a small stock of toxins and was operating equipment for the rapid breeding of various micro-organisms.

When MKSEARCH was created in 1963, the CIA director of plans, Richard Helms, noted in an internal memo that MKULTRA needed strengthening of 'structure and operational controls' and that CBR agent tests under 'simulated operational conditions' involved 'excessive risk to the Agency'. The MKSEARCH brief required the programme to: 'Develop, test and evaluate capabilities in the covert use of biological, chemical and radioactive material systems and techniques for producing predictable human behaviour and/or physiological changes in support of highly sensitive operational requirements.' Essentially there were two objectives: to develop CBR systems suitable for assassinations or sabotage and to investigate chemicals capable of altering human behaviour. The work was carried out by the CIA's Technical Services Division either alone or in conjunction with the US Army's Special Operations Division (SOD) at Ford Detrick.

The CIA-SOD link lasted for years on a highly secret but somewhat *ad hoc* basis. Many of the army scientists assumed that the Staff Support Group (the CIA cover name) was a branch of the military until they gradually came to realize who was actually involved. The cover name and funding citation were deliberately chosen to give a military appearance. Records are sparse and every effort was taken to keep paper work to a minimum by committing most details to memory. In later years, the CIA claimed to be unsure of what had actually been undertaken because many of those involved had left the agency.

One of the gadgets resulting from the CIA-SOD link was the M-1 projectile. It was described as 'capable of introducing materials through light clothing, subcutaneously, without pain' and of being fired from all sorts of devices — a pistol, a fountain pen, a walking stick and, as a forerunner of the Markov killing, an umbrella. The M-1 could carry incapacitating or lethal chemical, biological or toxin agents as well as lethal radioactive materials. Another product was a dart pistol which resembled a US Army .45 automatic but had a telescopic sight. It was said to be accurate up to 90 metres (100 yards) and completely silent. The gadgetry included more ingenious items. In the early 1960s the CIA toyed with various devices for assassinating Fidel Castro, including poisoned cigars, exploding shellfish (Castro is an ardent skin-diver), and booby-trapped radios and pistols. There was a plan to taint his shoes with a chemical which would make his hair fall out and psychologically 'emasculate' him with the loss of his beard. Another scheme involved tainting a cigar with a chemical to send him mad during one of his three-hour speeches. The chemical would probably have been BZ or another benzilate which would give the victim the appearance of a gradually increasing mental disorientation. As the effects may last up to 90 hours, the appearance of insanity would be medically consistent for longer than most drug-induced mental abnormalities. LSD can have similar but generally briefer, psychological effects and there was a plot to taint Castro's broadcasting studio with LSD or one of its potent chemical analogues. When President Nasser nationalized the Suez Canal in 1956, Britain's M16 evolved a plan to assassinate him by releasing nerve gas in a villa where he was staying. The nerve gas was stockpiled and the mission would have been the responsibility of the Special Air Service but it was cancelled by the Prime Minister, Anthony Eden.

The CIA's extensive interest in the benzilates and other psychoactive chemicals is well documented. From the early 1950s the agency embarked upon a programme of personality subversion triggered by the potentialities of the psychotomimetics and especially LSD. The basic idea was to manipulate or 'subvert' underlying motivations such as loyalty and cause an individual to act against his own or his country's interests without necessarily being aware of what he was actually doing. In some cases, the brief was to cause him to act without any recollection afterwards in a manner reminiscent of *The Manchurian Candidate*. The work seems to have been largely unsuccessful because of the complexity of the personality and the inability of psychologists to come to any genuine understanding of what is really involved in the fundamental relationships between morals and behaviour. Psychochemicals were also attractive because of their potential for inducing partial amnesia and were studied as a possible route to 'deprogramming' agents and contacts of sensitive information.

Both the US Army's Special Operations Division and the CIA held field experiments to determine the public's vulnerability to covert biological attacks. In the late 1960s SOD conducted simulated biological attacks on the New York subway system by using light bulbs filled with a dry powder containing the harmless agent *Bacillus subtilis*. Each bulb held nearly 88 trillion microorganisms mixed with charcoal to make the filling less noticeable when dropped from the train onto the road-bed. The movement of trains then distributed the agent down the line into other stations and trains. Data indicated that high concentrations of agent were dispersed over a fairly wide area although they tended to be greatest near the point of release. Persistence was for an hour or so. In other simulated attacks, BW devices were dropped through ventilating grilles from the street. Britain and the Soviet Union are thought to have conducted similar tests and there are suggestions that the Germans considered launching such an attack in Paris and London during World War II.

Although the documentation is more fragmentary, the CIA carried out its own covert biological warfare tests in New York in the early 1950s. One such test involved the dissemination of an unspecified chemical or biological agent (presumably a simulant) from a car with a modified exhaust pipe and from commercial plant sprayers concealed in suitcases. Another test in Florida may have used live whooping cough bacteria. According to researchers from the Church of Scientology who looked into local medical records of the time, the test may have been responsible for a sudden jump in the incidence of the disease after biological agents were released upwind from populated areas. Dozens of these tests have been conducted in America and the United Kingdom by intelligence agencies and the military. The purpose is usually to estimate the population's vulnerability to covert releases of biological agents from ships or aircraft, or by agents armed with the type of primitive but effective gadgetry used in the New York tests. The results indicate a high degree of vulnerability, but in many cases the researchers seem to have made some generous assumptions on the spread and aerobiological decay rates of the potential agents.

Today, BW would probably play a major role only in smaller wars or in the subversion of unstable regimes. Outbreaks of disease affecting crops or livestock could cause widespread social unrest. The essential prerequisite for destabilizing a regime and engineering a coup d'état is not to secure the active support of the majority of the populace but rather to ensure that the majority is either opposed to the ruling clique or indifferent to it. In using biological agents it would be necessary to make it look as though the outbreak was entirely natural by ensuring it was consistent with local disease patterns. The more primitive a country's economy and elementary its health services, the more effective covert BW would be.

The Terrorist Threat

In recent years the prospect of a terrorist nuclear weapon has been the subject of considerable concern and a great deal of highly colourful fiction. American authorities are said to have investigated a number of threats involving nuclear materials but, up to now, each has proved to be either a hoax or an attempt at blackmail through bluff. Although it might be possible to steal a military weapon or buy one on the black market, the real threat comes from the possibility that a terrorist group could acquire the essential materials and construct its own. This is not so much a question of having access to the necessary knowledge but rather obtaining sufficient uranium or plutonium and working it into a functional weapon.

All the information necessary to assemble a workable nuclear bomb with a yield up to 20 kilotons or so is available in scientific literature. Beyond this, the technicalities are probably too great but a 22-kiloton bomb devastated Nagasaki in 1945 and would do horrific damage in any large city today. The gun method used at Hiroshima would be the simplest design but it would not be suitable for plutonium because of the high chance of pre-initiation resulting in a very low or fizzile yield. An efficient gun-bomb would have to be based upon highly enriched uranium (well over 50 per cent uranium-235) and this imposes severe supply constraints upon clandestine groups. Stealing low-grade uranium might be fairly easy but the process of enrichment is expensive, extremely technical and is even beyond the resources of some middle-range industrial powers. A greater danger lies in the theft of weapons-grade uranium, but this would be more difficult than is usually suggested.

Similar problems exist with plutonium, but they are modified by the fact that weapons-grade material would not be essential. By highly competent design and careful engineering, a viable nuclear weapon could be made out of metallic or even non-metallic plutonium. Once again, obtaining the material is the chief hurdle that would have to be overcome; stealing reactor fuel rods and chemically separating the plutonium is almost certainly beyond the resources of most terrorist groups, but stockpiles of non-weapons-grade plutonium are becoming increasingly plentiful. A typical 1 billion watt commercial light-water reactor, for example, may produce up to 110 kilograms (250 pounds) of reactor-grade plutonium per year as a by-product. There is no clear definition of 'reactor-grade' plutonium; but generally, however, it means material containing 30 or 40 per cent of isotopes other than plutonium – 239. This makes potential weapon supplies of plutonium fairly abundant and protected only by the security of the storage depots — a security which varies widely from installation to installation.

Assuming a sufficient amount of plutonium can be obtained and processed into a suitable form — itself no easy task and fraught with dangers from radiation in all but the most modern facilities — the potential user would be forced to adopt the implosion method in constructing a bomb. The detonating circuit, detonators and wiring could be bought openly or stolen. The conventional explosive lenses which surround the fission sphere could be made by any competent chemist and their shaping and design could be undertaken by any trained physicist who had done preliminary research. The resultant bomb would certainly be less efficient than today's military weapons and would probably produce a yield far lower than predicted because of pre-initiation. There would even be a reasonable chance of a fizzile yield of around 1 kiloton instead of the 10 to 20 kilotons expected. But, as a terrorist exercise, even a relatively fizzile yield would be an extraordinary propaganda coup and cause appalling damage in any city.

There are other difficulties apart from getting sufficient nuclear material and actually constructing a workable nuclear device. To begin with, there is the very real problem of convincing the authorities that such a bomb had actually been built and put in place. It is not a question of having a government react to the threat — all such threats are taken seriously to begin with — but rather providing sufficiently convincing evidence that a genuine danger exists and it is not a bluff. If it is intended to use such a bomb for political objectives rather than as violence for its own sake, it would be critical that the authorities have no serious doubts as to the reality of what is involved. Secondly, however, there is the question of the demands themselves. If they are of the type associated with more traditional hijackings and kidnappings, then there is little need to go to the considerable trouble of making a nuclear bomb. But if the demands are totally unrealistic, the authorities are unlikely to meet them and may choose to gamble on locating and disarming the bomb or on being able to evacuate the threatened area. There is also the question of the deadline attached to the threat. If the time limit is too short, the authorities might not have sufficient time to assess their priorities or even meet the demands if — as is likely — they were difficult and extensive. A long deadline, on the other hand, would provide more time to locate the bomb and disarm it. None of these difficulties is insurmountable but, taking everything together, constructing and successfully using an illicit nuclear bomb to blackmail society is a much more remote possibility than is usually thought.

A more obvious danger is posed by nuclear waste. The likelihood of theft is small because nuclear waste is usually stored and transported in large metal and concrete drums which would be extremely difficult to remove. Instead, nuclear waste could be blown up in transit to spread radioactive contamination in the air. In 1980 the London *Observer* reported that a demonstrator carrying a dummy rocket-launcher had walked onto a railway platform where a train hauling nuclear waste was due to pass — according to a subsequent statement from British Rail, regulations did not forbid passengers carrying rocket-launchers from going onto station platforms. The sabotage of a reactor is another possibility and causing it to release radioactivity could contaminate a far wider area than simply blowing up a vehicle transporting nuclear material.

All things considered, however, chemical and biological terrorism is a more likely possibility than the use of clandestine nuclear or even radiological weapons. Palestinian groups are reported to have been stockpiling nerve gas for several years, and in 1976 an Austrian chemist was arrested for attempting to supply DFP (a relatively weak nerve agent) on the black market. Crude attempts at biological blackmail occurred in California in 1969, Chicago in 1972 and Hamburg in 1973. There are also vague rumours that the IRA has considered using biological agents in England. In 1980 French police raided a Germany Red Army safe house in Paris and found evidence of biological experimentation, literature on lethal diseases and traces of botulin in a rudimentary laboratory.

As the botulin bacteria itself is unsuitable for biological warfare, the experiments were almost certainly aimed at the separation and purification of botulin toxin. How the Red Army planned to use its botulin is not known but the substance can be used for tainting darts or bullets for assassinations, building crude explosive or aerosol BW bombs for terror attacks in cities and contaminating food stocks or water supplies. Biological agents might also be introduced into the ventilating systems of public buildings. Since the late 1940s this has been regarded as one of the most likely means of BW sabotage and a US Army test on the Pentagon found that between 0.5 and 1 litre (1 to 1½ pints) of biological agent inserted into a single air vent might be sufficient to contaminate the entire building. Filtration and purification systems in modern cities would prove a barrier to contaminating water supplies but it is conceivable that terrorists could succeed in using the advanced techniques of microencapsulation to coat BW particles with protective substances and make them resistant to such things as chlorine. However, this is probably beyond the capabilities of most terrorist groups. Crude biological bombs or aerosols would be fairly simple to design — although they would be highly dangerous to fill — and the release of lethal agents into the air would endanger large numbers of lives even if they did not cause major disaster or epidemics.

While chemical or biological terrorism might be less dramatic than the use of illicit nuclear munitions, the materials would be easier to obtain, work with, adapt into weapons, hide and transport to the point of use. Advanced equipment would be required but could be either stolen from universities and public health or industrial laboratories or openly bought from scientific supply houses. Illegal laboratories are not uncommon in today's world and are sometimes highly sophisticated, as are those involved in the processing of opium products into heroin or the manufacture of drugs like LSD and amphetamines. The necessary expertise is available, too. Many urban terrorist groups had their origins in university-centred sub-cultures and most members are either academically trained or maintain close links with the academic community. Groups like the IRA and various Palestinian organizations which did not arise from student radical politics in the 1960s and 1970s, have the advantage of being able to draw upon a pool of expertise provided through international cooperation in terrorism and the support of countries such as Libya.

Obtaining initial supplies of biological or toxin stocks would be quite easy. To make ricin all one needs is a supply of castor beans or seeds. Not only does the castor oil plant grow wild over much of the world but varieties are also frequently found in commercial nurseries where they are sold for gardens and homes. Any large-scale use of ricin would call for a great deal of starting material but even this might be obtained more easily than is commonly believed. Every manufacturer of castor oil takes the first step in ricin production when he separates the oil from the crude bean cake which carries the toxic constituent. Bribery or theft could provide abundant supplies. Microorganisms for breeding biological agents could be acquired from university, industrial or public health laboratories. The first American and Canadian supplies of *Clostridium botulinum*, for example, came from canning companies that had carried out considerable research into the dangers of botulin poisoning from tinned foods.

Filling large numbers of aerosols or biological bombs, however, is an entirely different matter and would require facilities beyond the capabilities of clandestine groups. The US Army's BW plant at Vigo, Indiana, had a maximum capability for producing 80 million cubic centimetres of slurried botulin filling per month which, once inserted into E-96 cluster bombs, would have been enough to cause 50 per cent casualties over an area of 4 square kilometres (1.6 square miles). But production at this level would have called for 12 90,000-litre (20,000-gallon) tanks and diverse supporting equipment. This would almost certainly exceed the resources of even the most wealthy and determined terrorist group anxious to avoid detection. Terrorist BW bombs would endanger life but would not cause the same level of massive casualties associated with biological warfare. An exception might be the use of a highly epidemic disease such as pneumonic plague, but the agent's unpredictability would make it impossible to be certain that it would spread decisively. Even if an epidemic did take hold, modern health services could probably contain it.

The value of biological warfare to the terrorist does not lie so much in the extent of the damage that would be inflicted but rather in the panic and psychological impact even a small success would be likely to engender. Just as one of the chief military virtues of strategic BW was the 'anxiety factor' and the public's fear of disease, use of biological warfare would have a propaganda value far beyond the likely casualty level. Plague would be especially effective in this regard even without epidemic spread because of its historical associations with the great pandemics of the Middle Ages. Even a small biological attack in a large city would cause the public to fear further outrages anywhere at any time and exploit the psychological value of BW's 'invisibility' first noticed by military researchers. This would be accentuated by the knowledge that it would be virtually impossible to locate and deactivate a number of biological weapons.

Lethal chemical weapons in terrorist hands would be potentially more destructive than biological agents. Compared with a fairly crude covert nuclear bomb, chemicals could well do more damage. Building a primitive nerve-gas bomb basically involves little more than filling a large tank with a chemical agent to surround an explosive core which, in turn, is linked to a detonating circuit. Producing the chemicals would impose safety and supply problems but none which could not be easily overcome compared with the difficulties of making a nuclear bomb. The precursor chemicals for nerve gas would have to be illicitly acquired, bought or produced from other compounds. The production processes for the major nerve gases are no longer a secret; quite the contrary, there is an abundant supply of recently declassified American and British patents in the open reference shelves of scientific libraries throughout the world and in the files of industrial companies. Binary routes would make things even easier and isopropyl alcohol, the sarin precursor, is commercially available.

Processes for sarin have been published for over 20 years but vary in complexity and suitability for safe production in clandestine laboratories. A very simple method for making small amounts of sarin and the other G-agents was published by the US Army in 1978. For sarin, m-nitrophenyl isopropyl methylphosphonate is mixed into a second solution containing tetraethylammonium fluoride and left to stand for a minute. Changes of chemical concentration, temperature and solvents can vary the rate of production and the quantity of the particular agent being made. Other published processes, more suitable for making large quantities of pure sarin, would be well within the capabilities of a competent chemist with access to reasonable laboratory facilities. A very simple US Army means of producing VM gas — only marginally less toxic than VX — is given in another declassified document. Sodium ethyl methylphosphonothiolate is put into a small body of water in which a preparation containing two other chemicals has already been added. The water then gradually converts to VM.

This process could also be used — as the US Army scientists took pains to point out — to contaminate small bodies of water such as 'ponds, lakes and streams' by adding the necessary three chemicals in the correct amounts. The water would turn into nerve gas and a point would be reached where fish and small wildlife would begin to die. The water would become increasingly more toxic until it threatened humans as well. The process can be neutralized through decontamination by chlorine and the document's wording illustrates the value of this method as a local pesticide. But there is no doubt that the procedure originated with warfare research and military land-wasting techniques. Adopted by malicious groups or individuals, this route to VM could be used either to make supplies of nerve gas or to poison an environment.

In Britain the existence of two patents for VX on the open reference shelves of the Patent Office in London provoked controversy when the *Sunday Times* revealed their presence in 1975. The two documents — as well as several others related to V-agent production — were immediately removed, an action that had to be legalized by retroactive parliamentary legislation some weeks later. By 1978 the Americans had published their V-agent specifications and copies were routinely filed in the Science Reference Library across the street from the Patent Office. When it was again pointed out that anybody could still obtain the information that Parliament had withdrawn three years earlier, the library reacted by removing a few more documents — but so haphazardly that it is still possible for anyone to go into the library and find a document on how to make VX, with no questions asked. Derwent's *Chemical Abstracts* apparently took the British government at its word and classified the two patent specifications for VX as insecticides within a section headed 'Basic Human Necessities'. Unlike the British, the Americans did not word their nerve gas patents as if they had insecticidal uses. On the contrary, each document states explicitly that the compounds 'are useful as chemical warfare agents'.

Forty kilos of VX is theoretically sufficient to kill everybody in the United Kingdom and just over twelve kilos would eliminate the entire population of California. In practice, such a disaster would be well beyond the capabilities of even the military, but crude chemical weapons in terrorist hands represent one of the gravest threats there is. The existence of scientific information on nerve gases and countless other warfare agents is not in itself threatening, but sooner or later the expertise will become available to those who wish to make war upon the social order. There is no current shortage of easily accessible lethal information or materials. A potato makes a perfectly good one-time silencer for a small-calibre pistol, certain commercial fertilizers provide a basis for cheap and effective explosives, an ordinary mousetrap is easily adapted into a trigger for a booby-trap bomb, and store-bought tampons provide excellent fuses for Molotov cocktails. These discoveries were not made by malevolent social dissidents but by armies teaching ingenuity and improvization to ordinary soldiers or providing specially trained commandos and agents with ready-to-hand weapons.

The illicit traffic in LSD and other psychoactive chemicals is the creation of the same university-orientated sub-culture that gave birth to groups such as Baader-Meinhof in Germany and the Weathermen in the United States. When the techniques of producing LSD first left the campus of the University of California at Berkeley to supply an increasing popular demand in the early 1960s, the foundations of a vastly profitable drug network were laid by enterprising chemistry students immersed in the pseudo-politics of rebellion and alternative culture. To a large extent this tradition has remained. When the British Operation Julie drug ring's chemist, Richard Kemp, was arrested he justified his activities as revolutionary and claimed to have donated large amounts of his vast profits to 'head' political groups. Julie detectives also found at least two individuals with links both to the LSD ring and to Baader-Meinhof, and there is reason to believe that the German terrorists were involved in some of the drug traffic in western Europe. The ingenuity, scientific expertise and organizational ability shown in the supply of psychoactive drugs is an excellent example of what can be done by amateur but proficient chemical enterpreneurs. Those motivated by nihilistic politics rather than profit could draw upon the same skills to produce lethal nerve gas agents and crude chemical weapons and then use them with great effect. Sooner or later malevolent groups may well begin to wage chemical and biological terrorism upon the existing order, in fact on the whole of humanity.

Notes

Note 1

A single electron volt is equal to 1.6021×10^{-19} joules or 1.6021×10^{-12} ergs (1 joule equals 10 million ergs) and 1 million electron volts equals 1.6021×10^{-13} joules. One electron volt is the energy acquired by an electron when it is accelerated by a potential difference of 1 volt.

Note 2

PRINCIPAL FUSION BOMB REACTIONS

REACTANTS	*PRODUCT NUCLEUS*	*PARTICLES*	*ENERGY (MeV)*[4]
A. Fission Trigger(s)	Fission Products	neutrons + photons[1]	200.0
B. *lithium-6 + neutron*	helium-4 + *tritium**	–	4.26
C. lithium-6 + deuterium	helium-4 + helium-4	–	22.0
D. *deuterium + tritium*	helium-4	neutron	17.6
E. (1) *deuterium + deuterium*[2]	helium-3*	neutron	3.26
(2) *deuterium + deuterium*	*tritium**	proton	4.04
(3) deuterium + deuterium	helium-4	–	24.0
F. *tritium + tritium*	helium-4	neutron + neutron	11.27
G. deuterium + helium-3	helium-4	proton	18.35
H. Secondary Fission[3]	Fission Products	neutrons + photons[1,3]	200.0

1. plus alpha and beta particles
2. of the three possible outcomes for any E reaction, 1 and 2 are equally likely but 3 is highly improbable
3. only occurs if uranium-238 is included in bomb design
4. total energy per reaction; in reaction D, for example, the product nucleus (helium-4) represents 3.5 MeV and the product neutron, 14.1 MeV. In reaction E(1), the neutron represents 2.44 MeV and the helium-3 nucleus, 0.82 MeV.

★ product nucleus available for secondary reactions

Italicised reactions are the most important

Note 3

A 1 kiloton nuclear explosion releases an energy of about 4.18×10^{12} joules and 1 joule equals 1 watt second. This is the same thing as 1161.1 megawatt hours or 48.3 days' output by a 1 megawatt power station. Since 1 megaton equals 1000 kilotons (or 1 million tons of TNT equivalents), a 1 megaton explosion is equal to 4.18×10^{15} joules. This is the same as 48,379.6 megawatt days or 132.5 megawatt years for a 1 megawatt power station.

Note 4

The formula for megaton equivalence (Mte) is:

(1) $Mte = nY^{\frac{2}{3}}$ for n warheads of Y megatons.

When the yield is greater than 1 megaton, most defence specialists suggest that the formula should be:

(2) $Mte = nY^{\frac{1}{2}}$

This is the formula followed in the text and tables.

Note 5
The probability of killing (Pk) any hardened target is a function of its lethal radius (LR), the attacking missile's overall reliability (R) and the warhead's accuracy in a circle of equal probability (CEP).
A missile's overall reliability R is the product of its component reliabilities for each stage of its operation. For a missile with five component reliabilities, for example, $R = r_1 r_2 r_3 r_4 r_5$.

Note 6
Measured in nautical miles, the CEP is a function of deviations from the intended impact point on the x and y coordinates. For a number of firings, the CEP is the median miss distance and, if the distribution is circular normal, 50 per cent of the warheads will land within the CEP and 99 per cent within an area four times the CEP.

Note 7
In nautical miles the lethal radius is calculated by:

(1) $LR = \frac{2.9Y^{\frac{1}{3}}}{H^{0.35}}$ for targets from 50 to 1000 psi or

(2) $LR = \frac{2.62Y^{\frac{1}{3}}}{H^{\frac{1}{3}}}$ for targets greater than 1000 psi where H is the blast resistance in psi and Y the yield in megatons.

Note 8
Once the LR, R and CEP are known, kill probability is calculated by:

$$Pk = R\left[1 - \tfrac{1}{2}\left(\frac{LR}{CEP}\right)^2\right]$$

By the same token, the chances of the target surviving an attack (Ps) by a single warhead is $Ps = (1 - Pk)$.

Note 9
If more than one warhead of the same Pk is fired at the target, the n-shot kill probability (ignoring fratricide) is

(1) $Pk_n = (1 - (Ps)^n)$ or $Pk_n = (1 - (1 - (Pk)^n)$.

But if the attacking warheads have different kill probabilities, then:

(2) $Pk_m = (1 - ((Ps_1)^{n1} (Ps_2)^{n2} \ldots (Ps_m)^{nm}))$ for m different survival probabilities of n warheads each.

Note 10
Formula 9(1) works for MIRVed warheads only so long as no two warheads are sent to the same target from the same launcher, i.e. only if the MIRVs are cross-targeted. When MIRVed warheads are not cross-targeted, the missile's overall reliability R is broken into two separate reliabilities, R_a and R_b. The first is the product of the component reliabilities from launch through warhead separation and the second, the product of individual reliabilities for the warheads through flight, re-entry and detonation. In this case n-shot kill probability is:

$$Pk_n = R_a (1-(1-R_b\left[1 - \tfrac{1}{2}\left(\frac{LR}{CEP}\right)^2\right]^n))$$

Note 11
A measure of the comparative effectiveness of different warheads against hard targets can be obtained through the lethality or K formula:

(1) $K = \frac{Y^{\frac{2}{3}}}{CEP^{2}}$. The larger the number, the more lethal the warhead. A single-shot kill probability (Pk) can be found according to:

(2) $Pk = R(1 - e^{[(-f(H))K]}$

The variable f(H) is a function of the hardness of the target. One frequently used model is:

(3) $$f(H) = \frac{In2}{(0.068H - (0.23(H^{\frac{1}{2}} + 0.19)))^{\frac{2}{3}}}$$

If Kt is the total lethality of a missile force attacking n separate targets (of the same hardness), the expected proportion of successes (S) is:

(4) $$S = 1 - \left[e^{-f(H)(Kt/n)}\right]$$

Note 12
One rad is equal to a radiation dosage of 100 ergs (6.25×10^{7} MeV) in tissue and one rem equals the dosage from any radiation which produces a biological effect in man equivalent to one rad of X-rays. The erg is a measuring unit of work and around 140 ergs are involved in lifting a pin about 1 centimetre.

Note 13
COMPARATIVE TOXICITIES

LETHAL

AGENT		*NATURE*	*SOURCE*	*APPROXIMATE DOSAGE*	
Botulinus toxin A (X)		p	Bacteria	0.00003	
Tetanus toxin		p	Bacteria	0.001	
Ricin (*W*)		p	Plant	0.02	
Palytoxin		np	Coelenterate	0.15	
Crotalus toxin		p	Rattlesnake	0.2	
Diphtheria toxin		p	Bacteria	0.3	MLD (minimum lethal dosage) in micrograms per kilogram
Cobra neurotoxin	*TOXINS*	p	Cobra	0.3	
Batrachotoxin		np	Frog	2.0	
Kokoi toxin		np	Frog	2.7	
Tetrodotoxin/Tarichatoxin		np	Fish/Newt	8.0	
Saxitoxin (*TZ*)		np	Dinoflagellate	9.0	
Bufotoxin		np	Toad	390.0	
Curare		np	Plant	500.0	
Strychnine		np	Plant	500.0	
VX		sc	sc	±7.5	
GD (*soman*)		sc	sc	50.0	LD50 in micrograms per kilogram
GB (*sarin*)	*NERVE GAS*	sc	sc	63.0	
GA (*tabun*)		sc	sc	150.0	
GF		sc	sc	400.0	

INCAPACITANT

AGENT	*NATURE*	*SOURCE*	*APPROXIMATE DOSAGE*	
Staphylococcal enterotoxin B (PG)	p	Bacteria	0.00004	
LSD25*	c	Alkaloid	0.002	
ALD52 (l-acetyl LSD)	c	Alkaloid	0.002	
BZ	sc	sc	0.0024	ED50 in milligrams per kilogram
MLD41 (1-methyl LSD)	c	Alkaloid	0.006	
DMHP	sc	sc	0.06	
STP*	sc	sc	0.07	
LBJ (N-methyl ester)*	sc	sc	0.1	
△l-THC (△l-tetrahydrocannabinol)*	sc	sc	0.5	

Note:
Italicized entries have been developed and adopted as warfare agents. Agents studied or adopted for covert use are not included.

Key:
p = protein
np = non-protein
c = chemical
sc = synthetic chemical
* = drug of abuse

Note 14

Much of the myth of biological warfare comes from interpreting toxicities literally. The fact that 1 ounce of botulin toxin A is more than enough to kill every person on earth is a statistic with immediate impact. But toxicities are highly dependent upon the nature of the experiment; the agent may be administered orally or through inhalation. Again it may be injected intravenously or into the muscles. Toxicities often vary widely between the various routes.

Toxicities are also likely to vary widely with the type of test animal and it is not always certain that any result is valid for humans. Thus if 0.00003 micrograms of botulin toxin A per kilogram of body weight is the minimum lethal injected dose for mice, and an average person weighs 70 kilograms, 0.0021 micrograms is the minimum lethal human dosage — but only if botulin has the same activity in men as it does in mice.

Taking things a step further, it would take 8.3307 grams (0.2938 ounces) of botulin to depopulate the planet — in 1975, the world's population was estimated at 3,967,000,000 — but only so long as everybody lined up for their injections and received exactly the right amount. In theory, spreading a few kilograms of botulin evenly over the earth would be enough to threaten the world's population through inhalation and ingestion but the practical problems of weapon inefficiency and aerobiological decay make reality dramatically different.

On the basis of the US Army's figures for the 4-pound, E-96 cluster bomblet, it would take 130,863,000 tons of BW filling charged with botulin toxin A to cause 50 per cent casualties in the American population. The attacker would have to blanket the United States with 551,749,000,000 bomblets and — given a plant with Vigo's capacity — spend 183,916 years making the BW filling at full production.

Note 15

SELECTED CHEMICAL AGENTS

	NAME	*CHEMICAL NAME*
Harassing agents	CA	α-bromobenzyl cyanide
	CN	σ-chloroacetophenone
	CS*	2-chlorobenzalmalononitrile

	NAME	*CHEMICAL NAME*
	AC*	hydrogen cyanide
	CK*	cyanogen chloride
	CG (phosgene)*	carbonyl chloride
	H (mustard gas)*	bis (2-chloroethyl) sulphide
	Q (sequimustard)	1,2-bis (2-chloroethylthio) ethane
	T2 Toxin	4β,15-diacetoxy-8α-3-methylbutyryloxy-12, 13-epoxytricothec-9-en-3α-ol
	GA (tabun)*	O-ethyl N, N-dimethylphosphoroamidocyanidate
G-agents	GB (sarin)*	isopropyl methylphosphonofluoridate
	GD (soman)*	1,2,2-trimethylpropyl methylphosphonofluoridate
	GF	cyclohexyl methylphosphonofluoridate
	VE	O-ethyl S-(2-diethylaminoethyl) ethylphosphonothiolate
V-agents	VM	O-ethyl S-(2-diethylaminoethyl) methylphosphonothiolate
	VS	O-ethyl S-(2-diisopropylaminoethyl) ethylphosphonothiolate
	VX*	O-ethyl S-(2-diisopropylaminoethyl) methylphosphonothiolate
	LSD25	lysergic acid diethylamide
	ALD52	l-acetyl LSD
	MLD41	1-methyl LSD
	BZ*	3-quidnuclidinyl benzilate
	DHMP (EA1476)	dimethylheptylpyran
	'STP'	2,5-dimethoxy-4-methylamphetamine
	'LBJ'	N-methyl-3-piperidyl benzilate hydrochloride
	$\triangle$l-THC	Delta l-tetrahydrocannabinol
	Mescaline Analog	3,4,5-dimethyoxy-α-methylphenethylamine
	Bufotenine	5-hydroxy-N, N-dimethyltryptamine
	Psilocybin	3-2-dimethylaminoethylindol-4-yl phosphate
	2,4-D	2,4-dichlorophenoxyacetic acid
anti-plant	2,4,5-T	2,4,5-trichlorophenoxyacetic acid
agents	picloram	4-amino-3,5, 6-trichloropicolinic acid
	cacodylic acid	dimethylarsinic acid

* Deployed or believed to be deployed as warfare agent.

Note 16
Chemicals in this group include O-n-butyl methylphosphonofluoridothioate, O-isopropyl methylphosphonofluoridothioate and methylphosphonothoic difluoride.

Note 17
Its full chemical name is 3-trimethylammoniopropyl methylphosphonofluoridate iodide.

Note 18
1-(N,N-dimethyl-N-(3-hydroxy)-propylammonio)-10-(N-(3-dimethylcarbamoxy-α-picolinyl)-N,N-dimethylammonio)-decane dibromide.

Note 19
1, 10-bis
(N-(3-dimethylcarbamoxy-α-picolyl)-N,N-dimethylammonio)-decane-2,9-dione dibromide.

Index